the essential guide to chinese american celebrations and culture

福 祿 壽

fortune   prosperity   longevity

HarperResource
An Imprint of HarperCollins*Publishers*

rosemary gong

good luck life

The author gratefully acknowledges permission to use excerpts from the following books and song lyrics: From *The Moon Year: A Record of Chinese Customs and Festivals* by Juliet Bredon and Igor Mitrophanow. Copyright © 1927 [1982] by Juliet Bredon and Igor Mitrophanow. Reprinted by permission of Oxford University Press (China) Ltd. All rights reserved. From *The Woman Warrior: Memoirs of a Girlhood Among Ghosts* by Maxine Hong Kingston. Copyright © 1976 by Maxine Hong Kingston. Reprinted by permission of Alfred A. Knopf. Lyrics from "Sun and Moon." From the Theatrical Musical Play *Miss Saigon* by Alain Boublil and Claude-Michel Schonberg. Music: Claude-Michel Schonberg. Lyrics: Alain Boublil and Richard Maltby, Jr. Copyright © by Alain Boublil Music Ltd. (ASCAP). From *The Moon Pearl* by Ruthanne Lum McCunn. Copyright © 2000 by Ruthanne Lum McCunn. Reprinted by permission of Beacon Press, Boston. From *China Men* by Maxine Hong Kingston. Copyright © 1980 by Maxine Hong Kingston. Reprinted by permission of Alfred A. Knopf. From *Wild Ginger: A Novel* by Anchee Min. Copyright © 2002 by Anchee Min. Reprinted by permission of Houghton Mifflin Company. All rights reserved. From *The Dim Sum of All Things* by Kim Wong Keltner. Copyright © 2004 by Kim Wong Keltner. Reprinted by permission of HarperCollins Publishers.

HarperCollins books may be purchased for educational, business, or sales promotional use. For information please write: Special Markets Department, HarperCollins Publishers Inc., 10 East 53rd Street, New York, New York 10022.

Designed by Judith Stagnitto Abbate/Abbate Design

Library of Congress Cataloging-in-Publication Data

Gong, Rosemary.
    Good luck life : the essential guide to Chinese
American celebrations and culture / Rosemary Gong.
        p. cm.
    Includes bibliographical references and index.
    ISBN 0-06-073536-8
        1. Chinese Americans—Social life and customs.
    2. Chinese Americans—Rites and ceremonies.   3.
    Holidays—United States.   4. Special days—United
    States.   5. United States—Social life and customs.
    I. Title.

    E184.C5G66   2005
    394.2'089'951073—dc22

2004054187

05 06 07 08 09 WBC/RRD 10 9 8 7 6 5

In loving memory of my grandparents, Low Hop Yee and Gong Yuen Tim, who gave me the gift of two worlds.

And to my parents, Tim and Mary Anna Gong, who taught me how to make my own happiness.

contents

Foreword / *ix*

Preface / *xiii*

The Chinese Calendar / *xv*

part I: ***annual chinese holidays*** / *1*

Chapter 1: Chinese New Year / *3*

Chapter 2: Qing Ming—Clear Brightness Festival / *43*

Chapter 3: Dragon Boat Festival / *59*

Chapter 4: Double Seventh Day / *77*

Chapter 5: Hungry Ghosts Festival / *85*

Chapter 6: Mid-Autumn Festival / *93*

Chapter 7: Chong Yang—Double Ninth Day / *107*

***part II: chinese special occasions* / 117**

Chapter 8: Weddings / **119**

Chapter 9: Red Egg and Ginger Party to Celebrate New Babies / **151**

Chapter 10: Big Birthdays / **169**

Chapter 11: Funerals / **189**

Chapter 12: Table Etiquette and Other Delicacies / **215**

Glossary / **235**

Bibliography / **251**

Acknowledgments / **255**

Index / **261**

contents / *viii*

foreword by Martin Yan

I was five years old, maybe six. It had been rain-ing all day, but the floor of our tiny kitchen felt familiar and warm. From my favorite spot under the kitchen table, I sat quietly watching my mother as she labored in front of our family's ancient built-in cast iron wok. My mom's kitchen was perhaps the root of the *Yan Can Cook* show, and I had the best seat in the house!

My mother, a short, wiry woman, looked even smaller next to the huge wok. If you think that I am quick around the chopping block, you should see my mother at work. Mom had this unique talent of making simple, everyday dishes absolutely delicious, and she could do it in the blink of an eye. In no time at all, that wonderfully comforting aroma of our dinner would fill the air. Could mom be making my favorite salted fish with steamed pork patties? Or perhaps tofu soup with fresh watercress and sweet dates? And what about velvety smooth steamed eggs with dried shrimp?

From my vantage point under the table, I noticed a small plaque at the corner of our kitchen. In front of it was a little urn of ashes with a few protruding joss sticks. From time to time, I noticed that my mother would place little cups of rice wine in front of it as an offering. "That's for our Kitchen God," she would say. "The Kitchen God protects us from all the bad things that can happen in the kitchen." "Like swallowing my watermelon seeds?" I asked.

In addition to being an excellent cook, my mother was also the best motivator. She knew just what to say to get me to do all my tasks around the house. And when I didn't finish all the rice in my bowl, instead of scolding me, she would remind me that every single grain of rice left behind in my bowl would be a pock mark on the face of my future bride. Needless to say, I made sure that I finished every grain of rice from that time on. What can I say—my mom knew her rice as well as she knew her son!

Yet, I never completely understood the explanation about our Kitchen God, or a few of the other odd things we did around the house such as making offerings to the hungry ghosts or never sticking my chopsticks straight into my bowl of rice. After a few more attempts to get to the bottom of the Kitchen God story, I gave up. I simply accepted him (or her) as a member of our family.

So many of our Chinese traditions and rituals have become an integral part of our daily lives. So much so that we sometimes take them for granted, without questioning their reasons or their origins. Like our shadows, we accept them as a part of us, and like our shadows, they follow us no matter how far away we find ourselves from China.

In my "Chinatowns" series, I had the privilege to visit many Chinese communities on different continents. I was amazed to find that despite the tremendous diversity in each of these overseas Chinese enclaves, the common bond of our Chinese heritage remains strong, and it is celebrated at every opportunity. Many Chinese immigrants arrived at their newly adopted countries with little more than the clothes on their backs, but they actually brought much more than that. They brought with them their history and culture, and lucky for us, they also brought with them all their great recipes!

What I find heartwarming is that our Chinese heritage is not lost over time. Children and grandchildren of immigrants, second- and third-generation Chinese Americans, Chinese Australians, and Chinese Canadians are showing a tremendous thirst to learn all they can about their Chinese heritage. The fact that they are more removed from their Chinese heritage only heightens their curiosity—they do not take any of these customs and rituals for granted. They dare to ask those questions that we, who have become accustomed to living with our shadows, have so often neglected.

Rosemary Gong is one of those who dares to ask the all too familiar questions. Why do we like the numbers three and eight but avoid the number four? Why do traditional Chinese families serve only vegetarian meals on the first day of the Chinese New Year? Instead of living with vague notions and then passing them on, Rosemary sets out to seek answers. Thanks to her efforts, *Good Luck Life* is a cultural treasure chest. It's a joy to reach in and come up with a nugget of history and folklore. What a great conversation starter it would make at any family dinner table! Going through these pages makes me realize just how much I thought I knew about my own culture, but really didn't, or at least, not completely. *Good Luck Life* paints vivid details of all the shades and colors that make up that shadow which is our Chinese heritage. Enjoy it with your entire family—it's a wonderful legacy to pass on to future generations.

—**Martin Yan**
Host, *Yan Can Cook*

preface

Ten years ago, if a Chinese fortune-teller had predicted that I would one day write a book about Chinese American customs, I would have been skeptical, because there were no signs for such a vision. I was living in the world of advertising—sandwiched between conference calls and ad deadlines. I barely made it home for Chinese New Year's Eve dinner. I had grown up in a small Central Valley town in California far away from Chinese school. In fact, my three siblings and I were the only Chinese in our school. If I had counted the words in my Chinese vocabulary, they would have totaled around twenty-nine and many weren't worth repeating. I couldn't tell the difference between a lotus and a peony.

As time passed, I began to wonder: Why do we Chinese visit the cemetery twice a year? What's the significance of bowing three times? How much money should go into a red envelope? What's wrong with giving all-white cut flowers as a hostess gift? What's the obsession with the color red?

I posed these questions to several wise aunties and uncles, and they generously shared their wisdom with me on the old Chinese ways. Quickly thereafter, I made the San Francisco Chinatown Public Library my second home. As if I were panning for gold, the nuggets of information began to surface, and I was excited to share my newfound knowledge.

This book is a culmination of my efforts to acquaint myself with my cultural heritage. It details historical facts, legends, common practices, and foods surrounding the Chinese holidays, including the Chinese New Year, Clear Brightness, Dragon Boat, and Moon festivals. Celebrations of life's milestones are also described—the Chinese wedding, the Red Egg and Ginger Party to welcome a new baby, significant birthdays, and the inevitable funeral. In each instance, I've attempted to make the connection between how and why the Chinese do what they do.

Good Luck Life also contains old village recipes for celebratory foods, reference guides to assist in building your own plans of action, decorum for different social situations, Chinese table etiquette for dining with confidence, and "do and don't" lists from wizen Auntie Lao, who recounts ancient Chinese beliefs and superstitions.

At the end of my quest, I found that the Chinese ways provided a life of abundance that overflowed with family and friends. *Good Luck Life* is a map to invite good fortune, practice honor, and deflect evil. The details can be adjusted and altered to embrace your own family's regional method of practice. On new soil, the old customs have evolved and new rituals have emerged, yet all are born from the same golden vein. With hope and generosity of spirit, *Good Luck Life* holds the promise to deliver fortune, prosperity, longevity, wealth, and health. May it serve you well.

農 曆

the chinese calendar

The Chinese calendar is a combined lunar-solar calendar that complies with the position of the sun and the phases of the moon. A regular year in the Chinese calendar has 12 months and 353, 354, or 355 days. But every 2 to 3 years, a leap year occurs that has 13 months and 383, 384, or 385 days.

The lunar calendar is based on the appearances of the moon. A lunar year has 12 months with 29 or 30 days in each month. However, China was a traditional agrarian society and therefore needed to follow the seasonal solar calendar for planting and harvesting. Consisting of four seasons, and a total of 365 days that are divided into twenty-four 15-day periods, the solar calendar is determined by the sun's longitude during the vernal equinox, summer solstice, autumnal equinox, and winter solstice.

An easy way to anticipate the first day of Chinese New Year is to identify when the new moon appears after the winter solstice, which typically falls between January 19 and February 23 (see page 40 for Chinese New Year dates through 2020).

The Chinese also name the years in each sixty-year cycle by combining a "celestial stem" of names (*jia, yi, bing, ding, wu, ji, geng, xin, ren, gui*) that have no English equivalent, and an "earthly branch" based on the Chinese astrological zodiac of rat, ox, tiger, rabbit, dragon, snake, horse, sheep, monkey, rooster, dog, and boar.

part one

annual chinese holidays

new   year

新年

Ask me about Chinese New Year and I think "cow tail" cookies. I think deafening firecrackers and red embroidered jackets. I think abundant trays of togetherness, whole white chicken, whole steamed fish. I think pairs. Pairs of tangerines, pairs of red envelopes, a pair of chopsticks. When I think of Chinese New Year, I think of Po Po and Gung, the perfect pair.

Chinese New Year is a time of new beginnings and intentions. Families sit down to feast on foods of good fortune once the clutter of the home, finances, and even the mind is cleared for a time of reflection, recognition, and renewal.

Traditionally known as the Spring Festival, which coincides with the seasonal farming calendar of the Chinese Almanac, Chinese New Year marks a fifteen-day celebration beginning on the first lunar new moon of the year and ending on the full moon. It usually falls between January 19 and February 23. Considered the most significant of holidays, the New Year integrates the themes of family, friends, home, and food. It's a time to put resolution and respect to practice and seek fortune, prosperity, longevity, happiness, and health.

The days leading up to Chinese New Year are fraught with flurry. Chinatown shoppers move to the rhythm of rustling pink plastic bags. Sidewalk vendors multiply with displays of seasonal flowers and blossoms, pallets of fresh fruits, and lively catches of the day. In preparation for the lunar New Year, the family readies itself by tossing out the old and welcoming in the new. The countdown begins with a chronological order of activities beginning with the Kitchen God ritual and moving on to the practices of settling old debts, readying the home, buying new clothes, and feasting to the family's content.

the kitchen god

About a week prior to the lunar New Year, on the twenty-third or twenty-fourth day of the twelfth lunar month, the Kitchen God, the most important domestic deity, is transported to the Jade Emperor, the ruler of the heavens, to report on the family's behavior from the previous year. The Kitchen God is represented by a paper image and is hung throughout the year near the family's stove. Long considered the soul of a Chinese family, the stove is where all is seen and heard. To encourage a good report, families smear the Kitchen God's mouth with honey or molasses, to sweeten his tongue. They remove his image from the stove and then burn it to send his spirit to the heavens. Some families offer spirit money during the deity's burning and even dip him in liquor to produce a bright flambé. When New Year's Eve arrives, a new Kitchen God is posted to replace the old one for another year of observation.

Today, many spiritual supply stores offer Kitchen Gods that are vertical wooden plaques painted red with gold Chinese calligraphy in addition to the traditional paper ones. These versions are intended to be permanent fixtures in your kitchen and Kitchen God joss papers are sold separately for his annual burning.

the man who would be kitchen god

It's said that the Kitchen God was a mortal named Zhang, a wealthy farmer whose lands and rivers flowed with abundance. Grains flourished in his fields, fish filled his rivers, and herds of livestock grazed his land. But Zhang wanted more. He took a mistress who drove his devoted wife away from their home. In the couple's excessive indulgences, Zhang and his mistress exhausted all of his wealth, and soon the woman deserted him for another man. Zhang, left with nothing, became a homeless beggar with no hope or will to live. So weak from starvation was Zhang that he suddenly collapsed fully expecting to die. He awoke in a mist of fog, which turned out to be the smoke from the hearth of a warm kitchen that welcomed all who possessed empty stomachs. Noticing Zhang's poor state, the kitchen girl fed and revived him. Nourished and bound by deep gratitude, Zhang sought to thank the mistress of the house, who was about to enter the kitchen from the outside garden. As she approached the door, he saw the mistress through a window and recognized her as his wife! Distraught and desperate, Zhang jumped into the hearth just as she entered the room, and the flame of shame grew large. Although Zhang's wife urgently tried to douse the fire, Zhang's ashes flew to the heavens in a huge, single *pheew!*

Upon hearing of Zhang's story, the Jade Emperor declared Zhang to be the Kitchen God. The heavenly ruler proclaimed that one who lived and learned as Zhang earned the gift of all-knowing and all-seeing and would influence the heavens year in and year out.

past due

Before New Year's Day, old debts and past quarrels should be resolved for a new start. The Chinese prefer not to carry forth the burdens of the past. In olden days, debt collectors carried lanterns on New Year's Day to suggest it was still New Year's Eve, thus extending extra time to settle up.

> **If a little money is not spent, great money will not be realized.**
>
> **—Chinese proverb**

sweep away

Serious spring-cleaning activities are conducted in preparation for a brand-spanking new year. Every corner is swept. The kitchen is scrubbed. Linens are laundered. Shutters and windows are washed. A clean house represents a fresh start and symbolizes the cleaning out of old misfortune. Cleaning must be completed by New Year's Eve and *not* on New Year's Day to avoid the risk of sweeping away any New Year's luck.

It's also a time to clean out the weevils of your mind. Some call it an attitude adjustment. During the New Year season, optimism rules. In the old days, Chinatown's very young and old alike lifted spirits by off-loading their laziness in time for the New Year with the playful Cantonese rhyme: *"Maai-lan, maai-lan! Maai-doh-nien-saam-shap-maan!"* ("Selling laziness, selling laziness! Laziness for sale until New Year's Eve!") No wonder the Chinese can be considered industrious.

wishing well

New Year "lucky papers" are simple declarations of good wishes. The Chinese prefer *wishes* over *resolutions* because if wishes aren't granted, the gods can be blamed. Resolutions are left to your own devices, and who wants that? As the New Year approaches, the prophetic wishes are hung inside and outside the home. Typical wishes are written in black or gold Chinese calligraphy with auspicious words such as "happiness," "wealth," "prosperity," "longevity," and the ever-popular *fu* (or *fook*) character, meaning "fortune" or "blessings."

Be creative and personal with your wishes for the New Year. Some examples have included the words "always full" hung on the kitchen cupboard, for ample reserves of rice; "good health" taped on the bathroom mirror, for excellent checkups; and "unexpected money," for trips to the casino.

Another form of "lucky paper" wall hangings are spring couplets. They're typically a pair of long vertical red paper strips written in black or gold Chinese calligraphy. Each strip is written in complementary verse and expresses good wishes and well-being for the family or business. The strips are hung outside on either side of a home's front door, in a prominent place inside the house, at a business's main entrance, or even on public gateways.

Other popular wall hangings grace Chinese shops and newsstands throughout the New Year celebration. Posters with folk art images of children, harvest scenes, birds, flowers, and fish offer the promise of luck and fortune. In Taiwan, the pineapple is a popular paper ornament, as it connotes prosperity. For those who fear the bogeyman, Door Gods are posted by the entryway. Representing images of Tang Dynasty Emperor Taizong's generals, Qin Shu Bao and Wei Chi Jingde, these imposing protectors on paper stand watch at the door as they did for the emperor so that he could get a good night's sleep knowing he was well guarded from the demons of the night.

fu 福

The practice of hanging the *fu (fook)* good luck character originates from an old tale of the Ming dynasty. While the emperor was traveling through his territory one day, he noticed a poster insulting to his empress hanging on many of the doors. Highly agitated, he counteracted it by distributing *fu* papers to the doors of all the homes that didn't possess the negative poster. By dawn, all the homes that didn't display the *fu* symbol had been destroyed. *Aiiya!*

In the northern tradition, *fu* is often seen hung upside down to take on the connotation that luck has arrived, because the words for "upside down" and "to arrive" sound similar in the Mandarin language. But for the Cantonese, this is not the case, and *fook* is always hung right side up.

red fever

Have you noticed how obsessive the Chinese are about the color red? This cardinal color is omnipresent as it adorns nearly everything from lanterns, stationery, and Chinese calendars to tassels, slippers, and jackets. Even foods such as roasted and barbecued meats carry that popular crimson shade. Red is the color of choice for decor and clothing during special occasions but especially for Chinese New Year. Long considered the most auspicious of colors, red is a sign of good luck because of its symbolic association with fire, the sun, brightness, life energy (the yang of yin and yang), and the lifeblood that demons fear most.

spring flora

Because of the proximity of Chinese New Year to spring on the lunar calendar, homes are decorated with fresh flowers to commemorate the hope a new year and season will bring. Several types of flowers hold significance, and red is the most auspicious and favored color. A prosperity tree—usually a miniature citrus tree or simply bare twigs—often completes the home. The Chinese use red string to tie red envelopes and old coins to the branches as an offering to the God of Wealth for a prosperous year. The following chart identifies some popular floral varieties and their symbolic meanings:

Flower	Significance
Narcissus	Good fortune and prosperity
Camellia	Spring
Evergreen Pine and Cypress	Great mortality, vitality, and endurance
Peach	Long life
Buddha's Hand Citron	Happiness and longevity
Quince Branches and Blossoms	Flourishing splendor and prosperity
Kumquat	Gold and prosperity
Orchid	Love and fertility
Peony	Spring and wealth

Flower	Significance *(continued)*
Lotus	Summer and purity
Chrysanthemum	Autumn; flower of the recluse, hardiness
Plum Blossom	Winter; pure, honest friendship; perseverance

To Chinese elders, New Year is one of the few occasions when a floral gift is appropriate, as flowers are traditionally associated with death and funerals. White flowers are a definite no-no, since white is a color of mourning. When in doubt, always give a Chinese auntie fresh seasonal fruit, candy, or cookies when visiting.

In Chinatown: San Francisco's Chinese New Year Flower Fair

Once the ball has dropped in New York City's Times Square, San Francisco's Chinatown begins to bustle with anticipation for the Chinese New Year. I typically begin my New Year's activities by purchasing fresh flowers so that they will be in full bloom on the first day of New Year. My fail-safe timing corresponds to the annual flower fair on San Francisco's Grant Avenue. Traditionally held on the weekend before Chinese New Year, this street fair begins on Grant and Broadway and runs for several city blocks, including Pacific Avenue and Jackson Street. The Chinese New Year Flower Fair is filled with hundreds of booths featuring seasonal flowers, spring couplets, home decorations, and food snacks to ready your home or business for the biggest Chinese holiday of the year.

firecrackers

The *pa, pa-pa-pa, pa, POW* of exploding firecrackers is believed to promote a change in energy, deliver new beginnings, and provide protection from harm. Because the ear-deafening bangs scared mortals and animals alike, it was believed that frightening away demons like Nian, the evil spirit, would make way for a year of good health, prosperity, and happiness. Firecrackers, *pau jeun* (exploding bamboo), mean all things to the Chinese. They are burned at home and at work, for the sacred and the secular, in celebration and in joy.

Chinese firecrackers come in different shapes and sizes. The most common ones are 1½ inches tall and are bundled sixteen or fifty per pack. These small red tubes are braided together with string and wrapped in red transparent paper. Some popular brands are Black Cat, Red Devil, China Doll, Mighty Mite, and Zebra. Industrial-proof versions consist of multiple ropes of firecrackers crowned by a hexagonal box and can equal up to fifteen thousand sticks of fire power. Strings of firecrackers are suspended on long poles during major public events, and their heart-stopping thunder never fails to satisfy the crowd while sprinkling the streets with lucky red confetti.

chinese chic

To welcome the birth of the New Year, the Chinese celebrate with a completely new outfit, and often with a tinge of red. Traditionally, this was the only time of year the Chinese indulged in such luxuries, so shoppers sought the best and dressed anew from head to toe, including festive overcoats and fancy undies. Although dressing in all black or all white may be elegant in the West, to the Chinese it's considered funereal dressing. So, throw in a splash of red and put your best Chinese-fashion foot forward.

paying respects

Flowers, food, candles, and incense are offered to a family's ancestors at the home's altar. It's an act of respect to honor and unite the family with those of the previous generations who've passed on. Letting the family elders eat first is considered a Chinese duty, and this applies to the dead as well as the living. Before sitting down to dinner, traditional families set a serving of the New Year's meal, including wine and tea, at the altar. It's a way of giving thanks, as the Chinese believe the family's good fortune is directly related to the well-being of its forebears. Once the dead ancestors have "eaten" their fill, the family isn't shy about consuming their leftovers, as being Chinese also means being practical.

chinese new year's eve

Chinese New Year's Eve is a night of reunion to instill harmony. A family dinner brings hope for a New Year filled with all things good. Children are allowed to stay up well past normal bedtime with the elders. Their enthusiasm for staying awake, known as the longevity vigil, is an auspicious sign of their elders' life spans. Red candles illuminate the house so that bad luck can't wander into the corners during the long night. Finally, at midnight, the new lunar year is welcomed with hearty conviviality and boisterous red firecrackers. Lucky money is distributed to the youngsters (and left under their pillows), and the spring that brings forth renewal has arrived.

Here are two common Chinese New Year's greetings in both Cantonese and Pinyin/Mandarin:

Cantonese	Pinyin/Mandarin	English
Gung Hay Fat Choy	Gong Xi Fa Cai	Wishing you happiness and prosperity
Sun Nien Fai Lok	Xin Nian Kuai Le	Happy New Year

lucky money

When a child expresses New Year's wishes and blessings to the grown-ups in Chinese, it ripens the pockets for lucky money. Known as *lai see* in Cantonese and *hong bao* in Mandarin, red envelopes are given to young or unmarried children by elders. Money in even amounts is considered lucky; however, four, which sounds similar to the word for death, is very unlucky. One contemporary Chinese method of gauging what amount to give is based on the price of a candy bar. Red envelopes are typically given one per adult. Married couples usually give two envelopes. But in some regions in northern China, mothers are the sole givers of the lucky red money packets.

Giving and receiving lucky money signifies good luck for all. Those who give will in turn receive. A Chinese family's luck is passed along and entrusted to their children. Receivers accept graciously with a hearty thank-you—*doi jeh* in Cantonese or *duo xie* or *xie xie* in Mandarin—and often with three kneeling bows. Etiquette dictates that the envelope remain unopened until the giver and receiver leave each other's company. New Year's lucky money is intended to sustain the child from one year to the next, so saving is encouraged. The word for red, *hong*, also sounds like "vast" or "liberal" in Chinese, leading to the belief that money wrapped in red will multiply.

chow time

A satiated tummy is the ultimate sign of well-being and affection in the Chinese culture. In fact, a common greeting among family and friends is "Have you eaten yet?" Therefore, it comes as no surprise that Chinese New Year is celebrated with many symbolic and fortuitous foods to invite a propitious new beginning.

Signifying unity and harmony, the New Year's Eve dinner is considered the most important family ritual of the year. Sibling rivalries and conflicts are set aside. The family dinner honors both past and present generations. All are encouraged to feast, including the family's ancestral spirits, who are fed at the family altar (see "Paying Respects," page 14). If a family doesn't share a New Year's Eve dinner, it's said the family's love will grow cold.

chinese new year's eve dinner menu

The dinner menu for New Year's Eve varies according to a family's preferences and tastes. These are some of the most significant and popular dishes served:

- A specialty soup, such as bird's nest soup for youthfulness and long life or shark's fin soup for prosperity, which often follows an appetizer course.

- A "monk's" vegetarian dish called *jai choy* is the most significant New Year's dish because every one of its ingredients promises to deliver good fortune, prosperity, and longevity: Dried oysters denote good business; *fat choy* (sea moss that looks like long black hair) means prosperity; Chinese black mushrooms fulfill wishes from east to west; *fun see* (long, clear bean threads) offer longevity; lily buds send one hundred years of harmonious union; lotus seeds deliver the birth of sons; dried bean curd encourages plenty; while cloud ears and snow peas are tossed in for more luck and prosperity.

- Poultry is traditionally served whole, complete with feet and head when practical, because serving only certain pieces is considered "broken" by the Chinese.

- Long leafy greens such as Chinese broccoli *(gai lon)* are served whole to wish a long life for parents. Long Chinese string beans also bring longevity.

- Whole fish, or *yu*, sounding like the word for abundance, is often the last dish served. While eating, the family should be careful not to flip the fish over. The head and tail should remain intact to ensure a good start and finish and to avoid bad luck throughout the year. This stems from an old fishermen's superstition of equating flipping a fish with flipping a boat.

- Long-grain rice or noodles for long life rounds out the dinner.

A meal containing eight courses, not including rice, is considered fortuitous because in Chinese "eight" sounds like "to grow." Dishes can be served banquet style, as it was once dictated by the imperial court. A banquet begins with cold or hot appetizers, followed by soup, fried foods, dishes sweetened with honey or sugar, preserved foods, meats, and seafood, and concludes with fresh fruit—especially oranges or tangerines. No Western-styled sweet desserts are served in a traditional Chinese dinner.

Leftovers from the New Year's Eve dinner are deliberate to signify that abundance is to be carried forth into the New Year. Customs dictate that no animals be killed on the first day of the year or cooking be done, so leftovers become very convenient. All utensils and supplies should be clean and unbroken, including dishes. A chip means something is eating into one's fortune. Chopsticks should be the same length to represent harmony.

fortified with fortuity

In northern China, the making and sharing of *yuanbao*, boiled dumplings, is a New Year's tradition that fosters family togetherness and cooperation of spirit. These dumplings are commonly referred to as *jiaozi* throughout the year, but during New Year they're called *yuanbao*, as in the currency of old China.

Filled with minced pork, shrimp, Napa cabbage, garlic chives, and seasonings, dumplings are eaten at midnight. Families gather to make and then enjoy them with the anticipation of who will receive the lucky dumplings containing a fortuitous blessing hidden in the center. Because the number eight represents prosperity, lucky dumplings are prepared by inserting one of eight symbolic items (see the following list). Eight of each kind of lucky dumpling are created for a grand total of sixty-four lucky dumplings. These special dumplings are then mixed and cooked with the full batch of regular ones. While boiling, it's important that the dumplings not break open in the water, or, it's said, the family's wealth will flow out. Once the dumplings are served, fate determines what goodness the New Year will bring for each family member.

Because dumplings are shaped like ancient gold ingots, eating these "golden nuggets" confers prosperity at New Year. Here are some traditional items used to insert good fortune into lucky dumplings:

- Peanuts for a long life
- Candies for a sweet life
- Dimes for prosperity
- Red dates for the early arrival of good things
- Dried longan fruit, or "dragon's eye," for well-roundedness in achievement
- Glutinous rice cake for achieving heights and promotion
- Oranges for positive outcome
- Walnuts for peace

new year's nibbles

Special snacks and cookies that are particular to Chinese New Year are often shared and exchanged among family and friends. Seasonal fruits such as tangerines, oranges, and pomelos (Chinese grapefruit) are stacked high in stores and widely displayed at home. Melon seeds, sweetened fruits, and candies are intended to pad the New Year with wealth, long life, and family. Below are some common Chinese New Year's snacks and food gifts:

- The Tray of Togetherness is an eight-sided sectional platter that holds a variety of snacks such as preserved kumquats, plums, sweet melon pieces, sugared coconut slices, and red dates. The center section traditionally holds red-colored melon seeds, which symbolize a family's continued lineage and wealth. Auntie Lao says those who eat melon seeds will create a penny for each seed eaten.

- Chinese New Year's glutinous rice cake, known as *nian gao,* is revered much like the Christmas fruitcake. *Nian gao* is sweet, dense, and sticky after steaming for hours. Made of glutinous rice flour, the plain version contains only sugar, oil, and white sesame seeds. Fancier versions may contain yams, nuts, and red dates. *Nian gao* is widely available in Chinese bakeries and is an appropriate holiday gift when making New Year's visits. The ingredients are symbolic of long life, harmony, and wishes for many children. When trimmed with sprigs of pine or cypress, they add longevity. *Gao* sounds like "high" in Chinese, so it signifies reaching for the heights of a better life in the New Year.

- Tangerines (known as *gut jai*) sound like "luck" in Chinese, so they're amply displayed and eaten. Tangerines with an attached leaf indicate new life and are often set out in pairs with a red envelope of lucky money at a child's bedside.

- Oranges deliver sweetness and wealth. Pomelos, *look yau*, mean having continuous prosperity and status. The color of citrus adds yet another layer of richness. Fruit is offered either in pairs or in even numbers (but not in the number four).

- The Chinese New Year brings a variety of cookies for sharing with family and friends. These cookies often include ones in red and purple square tins. Nicknamed "love letters" or "egg roll" cookies, they are delicate thin yellow rolled cookies that look like scrolls. When my grandmother came to visit, she brought New Year's cookies stored in red tin coffee cans. My favorites were *ngow sing* (cow tails), three-inch braids of fried dough, and *gok jai,* which were small flaky pockets with a sugary coconut, peanut, and sesame seed filling. In keeping with tradition, many first-generation Chinese mothers of the 1930s did not record recipes but rather committed them to memory. This was a time when the fine art of measuring was done with rice bowls, hands and fingers, and the feel of the dough.

食 mrs. chan's ngow sing—cow tail cookies

makes about 16 dozen

When I was a schoolgirl, cow tail cookies were my favorite Chinese New Year's treat. Cow tails are braids of fried dough that hold a crunch as well as the fragrance of the oil they're fried in. This recipe is from our Fah Yuen (Hua Xian) ancestral district, compliments of Mrs. Linda Chan. I remember being told that Crisco was the secret ingredient that gave our cow tails a light, delicate crunch, unlike the jaw-breaking packaged ones available at the market. But I always wondered how they got Crisco in old China.

- ¼ cup sesame seeds
- 1 package (¼ ounce) active dry yeast
- 2¼ cups warm water
- 8 cups all-purpose flour
- 5 teaspoons salt
- ¾ cup vegetable shortening
- Vegetable oil for deep-frying

1. In a small sauté pan over medium-low heat, toast the sesame seeds and let cool. In a small bowl, dissolve the yeast in the warm water. In a large bowl, mix the flour, toasted sesame seeds, and salt together. Add the vegetable shortening and work it into the flour mixture with your hands. Add half of the yeast-water mixture to the flour mixture while kneading the dough, adding the second half of the water mixture once the first addition is completely mixed into the flour. Knead the dough briefly and form into a loaf measuring approximately 2½ inches in height. With a kitchen knife, cut the loaf into slices 1 inch thick by 2½ inches high. Cover the dough slices with wet kitchen towels to keep moist. Proceed by cutting the dough slices into individual sticks that measure approximately ¼ inch thick by 2½ inches long.

2. To form cow tail braids, take a stick of dough and roll it with the palm of your hand against a flat surface, such as a table or cutting board,

into a long, thin string of dough resembling a shoestring measuring 16 to 18 inches long and less than ⅛ inch thick. Be careful not to make the dough strings too thick, or they won't be as crunchy once fried. Bring the two ends of the dough string together. Hold the center loop of the dough string with your index finger while rolling the opposite end of the dough in one direction until the dough string twists up, bringing the two ends of the string together. Secure the cow tail by inserting the ends of the dough into the center loop and place on a baking sheet covered with wet kitchen towels until ready to fry. Repeat with the remaining dough.

3. In a frying vessel at least 2 inches deep with a 2-inch clearance, heat the vegetable oil to 325°F. Test the oil's heat by dropping in a morsel of dough. If the oil sizzles and the dough quickly rises to the surface, the oil is ready. Gently drop the cow tails into the hot oil. When the cow tails rise to the top, press them into the oil until they begin to harden. Lightly toss the cow tails in the oil with a long pair of chopsticks or a slotted spoon until golden brown. Remove the cow tails from the oil with a flat wire-mesh strainer and place on baking sheets lined with paper towels to cool. Let the cow tails sit uncovered overnight to retain their crunch. Store in a foil-lined airtight container for up to two weeks.

食 auntie peggy's gok jai cookies

makes about 8 dozen

These flaky pastry pouches contain a sweet filling of coconut, peanuts, and sesame seeds. They're a family favorite during Chinese New Year. Before frying, my po po would decorate the pouches with three red dots by using a chicken feather and red food coloring mixed with water. This recipe comes from my Auntie Peggy Chu and her girls, Jamie and Vickie, who converted the village recipe, which called for ingredients according to handfuls, brick sections, and the size of peanuts, into U.S. measurements.

1 cup unsalted roasted skinless peanuts
2 tablespoons sesame seeds
2½ cups shredded sweetened coconut
2½ cups sugar
1½ cups lard, at room temperature
5¾ cups all-purpose flour
2 cups water
Corn oil for deep-frying

1. To prepare the filling, toast the peanuts in a small sauté pan over medium-low heat and let cool. In the same sauté pan over medium-low heat, toast the sesame seeds and let cool. Finely chop the peanuts with a cleaver or a sharp knife. In a large mixing bowl, combine the coconut, sugar, chopped peanuts, and toasted sesame seeds. Drop ½ cup of the lard in small pieces into the filling ingredients. Use your hands to mix until all of the ingredients are evenly blended.

2. To make the short dough, combine ¾ cup of the flour and ¼ cup of the lard. Work the lard into the flour until the dough adheres and can be shaped into a ball. Test the dough by taking a small pinch of it with your fingers; if the dough crumbles, the dough is ready. If the dough sticks together, add a touch of flour. When the desired consistency has been reached, set the short dough aside.

3. For the pouch dough, combine the remaining 5 cups flour in a large mixing bowl. Add the remaining ¾ cup lard in small pieces and work it into the flour with your hands. Slowly add the water to the flour-and-lard mixture and knead lightly until the dough forms. Overkneading will harden the dough.

4. Using ½ at a time, roll the pouch dough into a 1-inch-thick snakelike tube. Cut the dough into 1½-inch-long pieces. Continue rolling the re-

maining dough into tubes and cutting into 1½-inch-long pieces until all of the dough is cut. Cover the dough pieces with moist cheesecloth.

5. To form the pouches, take one piece of pouch dough and flatten it into a small circle the size of a silver dollar. Place a pinch (a scant ⅛ teaspoon) of the short dough in the middle of the dough circle and wrap it into a ball. With a small rolling pin, roll the dough out and then use your fingers to roll the dough up like a crescent roll. Turn the dough vertically and roll with the rolling pin. Again, use your fingers to roll up the dough like a crescent roll. Form the dough into a small ball and use the rolling pin to roll it into a 3½-inch circle. Place 1 teaspoon of the coconut filling in the center and fold into a semicircle pouch. Pinch the edges tightly to close.

6. To make a decorative rope edge, pinch a small corner of the dough up with your thumb and index finger. Slide your thumb under the pinched edge and fold the top edge over the center of the pinched area while making a new pinch slightly above the first. Continue pinching and folding around the pouch.

7. Repeat with the remaining dough and coconut filling. Cover the folded pouches with moist cheesecloth.

8. In a wok or deep-frying vessel, heat the corn oil to medium-high (325–350°F). Test the oil by dropping in a pinch of pouch dough. If the oil bubbles rise around the dough, the oil is ready. Deep-fry the pouches until golden brown, turning frequently. Remove the pouches with a wire-mesh strainer or slotted spoon and drain on paper-towel-lined baking sheets. In between batches, use a flat mesh sieve to remove stray bits from the frying oil. Sweet *gok jai* cookies are best when eaten after they have cooled completely, preferably after sitting overnight. Store in airtight containers for up to two weeks.

chinese food decoder

The family's selection of Chinese New Year dishes is based on tradition handed down through generations. The stories and symbolism for eating specific dishes vary by region. In some cases, given the tonal quality of the Chinese language, some words can take on other meanings with a slightly different tone. For example, in Cantonese, green onion, or *choon*, sounds like the word for smart; fish, known as *yu*, sounds like the word for abundance. Combining foods can expand the concept further. Placing tangerines (fortune) and lychees (advantage) together has greater auspicious significance, while tangerines alone simply mean luck. The chart below lists some popular Chinese foods with defined meanings:

Food Item	Significance
Candied Coconut	Togetherness
Chinese Black Mushrooms	Wishes fulfilled from east to west
Fat Choy (Sea Moss)	Prosperity
Fish	Abundance
Green Onions	Smart
Kumquats	Gold
Long-Grain Rice	Long life
Long Noodles	Long life
Longans	Many sons
Lotus Seeds	Children, long continuous lineage
Lychee Nuts	Strong family tree
Meatballs	Happy reunion
Melon Seeds	Progeny, many sons

Food Item	Significance *(continued)*
Nian Gao *(New Year's Cake)*	Reaching soaring heights
Oranges	Gold
Oysters	Prosperous business
Peaches	Longevity
Peanuts	Long life
Pineapple	Prosperity
Pomegranates	Children
Red Dates	Early prosperity, all good things
Red Dates and Chestnuts (combined)	Early son
Red Dishes	Good luck
Sweet Dishes	Sweet life
Tangerines	Luck
Tangerines and Lychees (combined)	Auspicious
Tang Yuan *(Rice Flour Dumplings)*	Reunion

the fifteen days of chinese new year

From Chinese New Year's Day to the first full moon, a period that culminates with the Lantern Festival, various days are designated as "birthdays" for several earthly beings and life-sustaining crops by acknowledging their day of creation. The Chinese honor these beings by paying homage on their annual day. Other days are allotted for family visits and for recognizing the heavenly deities. Here's a guide for each day's traditional practices:

Chinese New Year's Season	Acknowledgment and Activity
1st Day	A day of rest and visiting.
	Respects paid to ancestors.
	Temple visits for giving thanks and having fortunes told.
	Vegetarian day signifying no killing on the first day of the New Year.
	Offerings made to the God of Happiness.
	Birthday of chickens.
2nd Day	Married women visit their parents and are given sugarcane and lettuce to take home for sweet blessings for their families.
	Birthday of dogs.
3rd Day	Day of leisure at home.
	Floors swept to make room for the new.
	Wedding day of the rats.
	Early bedtime for humans.
	Birthday of pigs.
4th Day	Dinner at the stroke of midnight to acknowledge the God of Wealth for a prosperous year.
	Earth visited by heavenly deities.
	Offerings of incense, food, and spirit money to the heavenly deities.
	Birthday of ducks and sheep.
5th Day	A look to the weather to gauge the New Year's fortune.
	A safe day to empty the trash.

Chinese New Year's Season	Acknowledgment and Activity *(continued)*
5th Day (continued)	Birthday of the Gods of the Five Directions.
	Birthday of cows.
6th Day	Birthday of horses.
7th Day	Adults eat *yu san,* a raw fish salad with lettuce that symbolizes an abundant and prosperous life.
	Birthday of humans.
8th Day	Birthday of rice.
9th Day	Birthday of fruits and vegetables.
10th Day	Birthday of grains.
15th Day	Lantern Festival.
	Lanterns hung for progeny.
	God of Destiny rises to fulfill desires, and the Goddess of the Sea is ready to accept wishes.
	Offerings are tossed to the river:

- Prosperity of oranges for husbands
- Fragrance of apples for wives
- Longans for progeny
- Red dates for all things good
- Pebbles for a house
- Coins for treasures

It is interesting to note that in Southeast Asia, yu san, raw fish salad, is enjoyed throughout the Chinese New Year season and widely served during the holiday's celebratory banquets.

lanterns by the moon

When the first full moon of the year rises on the fifteenth day of Chinese New Year, lanterns are hung to symbolize the light and warmth of spring. Extra lanterns are also an invitation for newborn children. The gods are out in full force, and festivities are set to honor the Sun God.

This is the day marriages are made on the moon. It's said that the old Moon Minister of Marriage matches baby girls and baby boys by encircling them for life with an enchanted red thread. Years later, when their paths cross on the night of the first full moon, it's a heaven-on-earth connection.

Years ago, peasants of southern China gathered to eat yams under the lanterns at midnight. Yams are sweet and the rich ocher of harvest, so it's believed that eating yams satisfies the soul and prevents it from longing to leave life and join the ancestors.

Lanterns are used as signposts for the ancestral spirits to guide them home for the lunar celebration. After the fifteenth day, the light leads them back to the otherworld.

a sweet solution

Tang yuan are sweet rice flour dumplings served in syrup and are the God of Fire's favorite food. This dish stems from the legend of Yuan Xiao, a young servant girl who was so homesick for her family, she was about to take her own life. Her friend Dong Fang, an imperial courtier, came to a solution. He'd heard that the Jade Emperor had commanded the God of Fire to destroy the city on the sixteenth day of the New Year. After some thought, Dong Fang persuaded Yuan Xiao to disguise herself as the God of Fire by wearing a red dress and to convince the emperor to order all households to cook *tang yuan* as an act of appeasement to the fiery destroyer. As the real God of Fire feasted on his favorite food, lanterns were hung and fireworks lit to give the impression that the city was burning. This satisfied the Jade Emperor's mandate and was cause for celebration. Subsequently, Yuan Xiao was reunited with her family, who came to the city for the festivities. The ingeniousness of Dong Fang laid the groundwork for resolution, and the legend of the Yuan Xiao Lantern Festival was born along with the practice of eating *tang yuan* during special occasions to symbolize reunion.

auntie lao says . . .

Growing up Chinese American means that there are all kinds of do's and don'ts, and Chinese New Year is no exception. Wizen Auntie Lao is everyone's wise auntie, a little Chinatown lady who's smart in the old ways of China and primed to ward off evil. Feel free to pick what works for you and leave the rest to destiny.

- Don't shampoo your hair on New Year's Day, or the year's good luck will be washed away.

- Don't sweep the floors on the first day, or good fortune will be swept away.

- Don't use sharp edges, knives, or scissors on New Year's Day, or prosperity will be cut away. Chinese hair salons can take the day off.

- Hold your tongue against cursing and gossiping to bring a year filled with peace.

- Breaking things will bring seven years of bad luck. A quick Cantonese remedy to this predicament is to say *"lok dai hoi fa,"* meaning "Fall to the floor and open with flowers." While in Mandarin, *"Sui sui ping an"* counteracts with peace for every year because "broken" also sounds like "peace."

- Wear jade jewelry to manifest good luck.

- Be Chinese fashion—forward by mixing red with pink.

- Dressing in all black or all white is a no-no.

chinese new year's parade

Chinatowns across America sparkle during Chinese New Year, and the hallmark of the festivity is the New Year's parade. This occasion calls for a place in the front row. The parade is a visual feast with decorated floats, lion dance troupes, costumed Chinese folk legends such as the Monkey King and the Eight Immortals, high school marching bands, and Miss Chinatown and her court. It's a citywide cultural affair, with state and city officials waving alongside elementary schoolchildren dressed as the designated animal of the year.

The oldest and largest Chinese New Year's parade in the United States is the San Francisco Chinese New Year's Parade. Since 1958, the parade has been under the direction of the Chinese Chamber of Commerce to complement the Miss Chinatown USA Pageant. This event alone attracts millions of spectators and television viewers annually and is held on the Saturday closest to the full moon that marks the end of the two-week New Year's celebration.

The parade's star attraction is a 201-foot-long golden dragon known as *Gum Lung*, who twirls through the streets aglow on a cloud of exploding firecrackers. A creature of the fire-breathing variety and custom-made by China's Foshan dragon masters, *Gum Lung* is decorated with the colors of the earth, fire, water, and wind. The dragon possesses a 6-foot-long head, and colored bulbs glow from head to tail. Silver rivets and white fur outline the dragon's scales. Constructed of bamboo and rattan, *Gum Lung* comes in twenty-nine segments and requires a small army of one hundred people to carry it through the streets. At the end of the parade route, the golden dragon is met with nearly a million firecrackers strung on a pole the height of a multistoried building. Exhilarated cries escape the crowd as thunderous firecrackers dance over a whirling *Gum Lung*, marking the parade's end and time for a late-night snack.

lion dancers

From January through April, lion dancers prance throughout Chinatown as a ritual to whisk demons away and to bring prosperity for Chinatown businesses. The Chinese have long considered lions as protectors. The lion dance consists of ornately costumed synchronized dancers and martial arts practitioners who perform to the rhythm of clanging cymbals and persistent gongs and drums. With flapping jaws and batting eyes, the lion prances and bows in pursuit of a decorated ball that represents a priceless pearl. In exchange for exorcising evil spirits, lucky money in red envelopes rewards the troupe for its services.

Many of these troupes are nonprofit martial arts schools and civic organizations, so the donations collected help support their tradition.

astrological animal years

Each Chinese New Year welcomes a year represented by an animal that is associated with the lunar calendar's twelve astrological animals. According to legend, when the Earth God was establishing all things on the planet, he held a race of animals to determine the Chinese calendar system. Only the first twelve animals to cross the finish line would be represented. The cat and rat were concerned about their size and strength in comparison to the other animals, so they partnered with the nearly blind ox, who needed guiding during the race. The ox struck a deal with the rat and cat to ride on his back and act as navigators. From the race's start, the ox easily held the lead. All appeared in order until the cunning rat pushed the cat off the ox's back while crossing a river. Then, just as the ox approached the finish line, the rat jumped off the ox's back and into the lead. The rat was declared the first animal to cross the finish line, followed by the ox, tiger, rabbit, dragon, snake, horse, sheep, monkey, rooster, dog, and boar. The cat was the thirteenth animal to finish and did not secure a place in the Chinese calendar. Since that time, cats and rats have been sworn enemies.

Select your Chinese astrological sign and persona on the pages that follow to help decipher the fortune of your present and future. It's worth noting that the dates represent the lunar year, so if you have a late January or early February birthday on the Gregorian calendar, you may be on the cusp of two animal signs. To double-check your sign, confirm which date Chinese New Year occurred in the year you were born.

Rat

Year of Birth:	1924, 1936, 1948, 1960, 1972, 1984, 1996, 2008
Positive Characteristics:	Imaginative, charming, generous, cunning, opportunistic
Negative Characteristics:	Short-tempered, critical, nervous, selfish
Possible Careers:	Sales, critic, publicist, writer
Compatible Mate:	Monkey
Incompatible Mates:	Horse, rooster, sheep
Famous Rats:	Marlon Brando, Gwyneth Paltrow, Sean Penn

Ox

Year of Birth:	1925, 1937, 1949, 1961, 1973, 1985, 1997, 2009
Positive Characteristics:	Leader, conservative, methodical, dependable, honest, patient
Negative Characteristics:	Chauvinistic, self-centered, stubborn, humorless
Possible Careers:	Surgeon, military general, hairdresser
Compatible Mates:	Rooster, snake, rat
Incompatible Mates:	Tiger, sheep, dog
Famous Oxen:	Jack Nicholson, Bruce Springsteen, Vera Wang

Tiger

Year of Birth:	1926, 1938, 1950, 1962, 1974, 1986, 1998, 2010
Positive Characteristics:	Sensitive, rebellious, adventurous, energetic, passionate
Negative Characteristics:	Quick-tempered, restless, suspicious, petty, impulsive
Possible Careers:	Explorer, race car driver, supervisor
Compatible Mates:	Dog, horse, boar
Incompatible Mates:	Monkey, snake, ox
Famous Tigers:	Tom Cruise, Peter Jennings, Hello Kitty

Rabbit

Year of Birth:	1927, 1939, 1951, 1963, 1975, 1987, 1999, 2011
Positive Characteristics:	Pleasant, affectionate, gentle, artistic, sophisticated, cautious
Negative Characteristics:	Indifferent, temperamental, secretive, disloyal
Possible Careers:	Lawyer, diplomat, actor
Compatible Mates:	Sheep, boar, dog
Incompatible Mates:	Rooster, tiger, horse
Famous Rabbits:	Francis Ford Coppola, Jet Li, Brad Pitt

Dragon

Year of Birth:	1928, 1940, 1952, 1964, 1976, 1988, 2000, 2012
Positive Characteristics:	Vital, ambitious, daring, gifted, confident, forceful
Negative Characteristics:	Arrogant, abrasive, angry, demanding, unwilling to compromise
Possible Careers:	Politician, priest, artist
Compatible Mates:	Monkey, rat, snake
Incompatible Mate:	Dog
Famous Dragons:	David Ho, Maxine Hong Kingston, Amy Tan

Snake

Year of Birth:	1929, 1941, 1953, 1965, 1977, 1989, 2001, 2013
Positive Characteristics:	Wise, charming, romantic, intuitive, powerful
Negative Characteristics:	Manipulative, ruthless, vindictive, possessive, demanding
Possible Careers:	Philosopher, writer, psychiatrist
Compatible Mates:	Ox, rooster, dragon
Incompatible Mates:	Boar, tiger, monkey, horse
Famous Snakes:	Tony Blair, J. K. Rowling, Oprah Winfrey

Horse

Year of Birth:	1930, 1942, 1954, 1966, 1978, 1990, 2002, 2014
Positive Characteristics:	Hardworking, independent, intelligent, friendly, athletic
Negative Characteristics:	Egotistical, rebellious, indecisive, volatile, impatient
Possible Careers:	Adventurer, scientist, poet
Compatible Mates:	Tiger, dog, sheep
Incompatible Mates:	Rat, ox
Famous Horses:	Jackie Chan, Michael Moore, Barbra Streisand

Sheep

Year of Birth:	1931, 1943, 1955, 1967, 1979, 1991, 2003, 2015
Positive Characteristics:	Elegant, artistic, calm, generous, compassionate
Negative Characteristics:	Nervous, moody, pessimistic, shy, insecure
Possible Careers:	Actor, gardener, artist
Compatible Mates:	Tiger, horse, boar
Incompatible Mates:	Rat, ox, dog
Famous Sheep:	Bill Gates, Steve Jobs, Yo-Yo Ma

Monkey

Year of Birth:	1932, 1944, 1956, 1968, 1980, 1992, 2004, 2016
Positive Characteristics:	Intelligent, nimble, polite, ambitious, versatile, witty

Monkey (continued)

Negative Characteristics:	Vain, deceitful, selfish, jealous
Possible Careers:	Success in any field
Compatible Mates:	Dragon, rat
Incompatible Mates:	Tiger, snake
Famous Monkeys:	Tom Hanks, Michelle Kwan, George Lucas

Rooster

Year of Birth:	1933, 1945, 1957, 1969, 1981, 1993, 2005, 2017
Positive Characteristics:	Shrewd, articulate, honest, hardworking, meticulous
Negative Characteristics:	Grandiose, temperamental, relentless, combative
Possible Careers:	World traveler, restaurateur, public relations
Compatible Mates:	Snake, ox
Incompatible Mates:	Rabbit, rat, dog
Famous Roosters:	Edward Norton, Yoko Ono, Diane Sawyer

Dog

Year of Birth:	1934, 1946, 1958, 1970, 1982, 1994, 2006, 2018
Positive Characteristics:	Trustworthy, faithful, fair, protective, intelligent, brave
Negative Characteristics:	Stubborn, outspoken, critical, reserved
Possible Careers:	Business leader, activist, teacher

Compatible Mates:	Tiger, horse
Incompatible Mates:	Rooster, sheep, dragon
Famous Dogs:	George W. Bush, Bill Clinton, Donald Trump, Tim and Mary Gong

Boar

Year of Birth:	1935, 1947, 1959, 1971, 1983, 1995, 2007, 2019
Positive Characteristics:	Sincere, loyal, honest, resilient, courageous, generous
Negative Characteristics:	Materialistic, domineering, immovable
Possible Careers:	Producer, writer, lawyer
Compatible Mates:	Rabbit, sheep
Incompatible Mates:	Snake, monkey, boar
Famous Boars:	Hillary Clinton, Maya Lin, Arnold Schwarzenegger

springing forward by planning backward

Planning for Chinese New Year can be compared to a multiday seasonal festival, complete with gift-giving, decorating, family visits, and bountiful multiple-course banquets.

Here's how you can begin preparing before the smoke from the Kitchen God takes fire:

step one

Identify when Chinese New Year falls for the coming year and then plan backward. Chinese New Year's Day typically falls between January 19 and February 23, so an easy mnemonic device is to remember it falls sometime between Martin Luther King's Birthday and Valentine's Day. For easy reference, here are the dates for Chinese New Year's days through 2020:

Year	Chinese New Year's Day	Chinese Astrological Year
2005	February 9	Rooster
2006	January 29	Dog
2007	February 18	Boar
2008	February 7	Rat
2009	January 26	Ox
2010	February 14	Tiger
2011	February 3	Rabbit
2012	January 23	Dragon
2013	February 10	Snake
2014	January 31	Horse
2015	February 19	Sheep
2016	February 8	Monkey
2017	January 28	Rooster
2018	February 16	Dog
2019	February 5	Boar
2020	January 25	Rat

step two

Easily organize your Chinese New Year's activities according to the chart below for the countdown to Chinese New Year's Eve:

Activity	Suggested Timing
Shop for Chinese New Year's attire. (Remember the color of the season is red.)	1 to 2 weeks prior
Spring-cleaning of household.	1 to 2 weeks prior to Chinese New Year's Eve
Plan, shop, and prepare the foods and feasts of the New Year's season.	1 to 2 weeks prior—all the way up to Chinese New Year's Eve
Decorate with fresh flowers, spring couplets, and "lucky" wall hangings.	1 week prior to the day leading up to Chinese New Year's Eve
Fire up the Kitchen God to the Jade Emperor.	12th lunar month, 23rd/24th day (5 days prior to Chinese New Year's Day)
Prepare Chinese New Year's family dinner.	2 to 3 days prior to New Year's Eve
Repay old debts.	1 to 3 days prior to New Year's Eve
Stuff red envelopes with lucky money.	1 to 2 days prior to New Year's Eve
Kick back and eat leftovers; it's Chinese New Year!	Chinese New Year's Day

clear and pure　　　　　*bright*

chapter two / qing ming—clear brightness festival

When I think of Qing Ming, I think of an April-morning pilgrimage to visit the faces of my past. Souls that once slept by my side, now nowhere to be seen. The Chinese cemetery is a place unspoken. A piece of heaven. A slice of hell. A place of duty. Wash down. Feed mouths. Send money. The fragrance of burning incense, popping firecrackers, and freshly cut flowers permeates the air. Three bows of respect can influence fate and, at the very least, warm a cold heart.

It's traditional Chinese belief that a person's good fortune is directly linked to the happiness of one's ancestral spirits. The best time to secure the family's elders' joy is during the Clear Brightness Festival, also known as Qing Ming (pronounced "Ching Ming"). It's a springtime holiday around April 4–6, or 106 days following the winter solstice, that acknowledges the dead in a cemetery ritual. Qing Ming is a Chinese holiday that corresponds to the Gregorian calendar.

Clear Brightness originally coincided with the spring willow-planting festival of ancient China. It was an old version of the modern Arbor Day. The emperor would plant new cuttings annually on the imperial grounds. But as Confucian belief began to take root, the Han dynasty (206 B.C.E.–220 C.E.) established Qing Ming as a day for venerating ancestors to reflect the supreme Confucian virtue of familial loyalty.

In old China, Qing Ming was a time for the living to care for the dead. In return, the living would reap a successful planting season and a bountiful harvest. This custom stemmed from the belief that ancestors who dwelled in the otherworld possessed powers that ruled nature and destiny. Honoring the dead provided the living with a sense of peace and identity. Today, the festival of Qing Ming marks the devotion of the present generation to the generations of the past and provides a heightened sense of familial continuity for all future generations.

> *The ritual of venerating departed ancestors is also referred to as heng san in Cantonese, a phrase that literally means "walk mountain," because many Chinese cemeteries are nestled on hillsides according to feng shui principles.*

qing ming in four steps

The Clear Brightness Festival is observed with a cemetery ritual that nourishes and remembers ancestral spirits in four steps.

step one: tidy up

The winter storms often leave the ancestral grave site in disarray. The first step is to clear away the debris by pouring water on the headstone to wash away what nature has left behind. Remove withered leaves, old pine needles, cobwebs, and grass shavings. A gallon of water and old towels come in handy for this task.

step two: food and flower offerings

Lay fresh flowers once the grave site is readied for new spring offerings. Usually a family places a spring bouquet of flowers for each of their departed relatives.

Next display food offerings in front of the ancestral headstone. Families serve the food on Chinese porcelain or paper plates or in Styrofoam containers, pink bakery boxes, or even tinfoil turkey roasters for big eaters.

The food offerings in America varies according to family preferences. Common items are crispy-skinned roast pork, whole boiled chicken, steamed pork buns, golden sponge cake, custard tarts, oranges, apples, and steamed rice. Any and all types of food make an attractive offering for loved ones. Sometimes families feed their dead with their favorite dishes. I've even spotted a whole all-American apple pie on a grave site during Qing Ming.

Beverages typically served include tea and Chinese liquor or wine. Tea and wine cups are often placed at the top of the food offering nearest to the headstone. Remember to pour the liquids into the ground before leaving so that the ancestors can take full advantage of the drink.

> *Ancestral food offerings are made in even numbers to correspond with the yin principle relating to the clouds and heaven. Gods and deities receive odd-numbered offerings because they're considered yang.*

step three: spirit world offerings

The key ingredients for typical spirit offerings used during Qing Ming are widely available in Asian markets and communities. Ritual packages for the dead can be purchased from "otherworld spirit" shops in Chinatown. Statues of Chinese deities are a visual giveaway for identifying stores selling otherworld supplies. Ask for assembled packages for the cemetery ritual and you'll be handed a bag with a few odds and ends that usually costs less than five dollars. The *heng san* ritualistic kit contains a pair of red candles, a handful of incense sticks, otherworld spirit money, and sheets of brown joss paper (see "Qing Ming Checklist," page 55). Each kit serves one grave site. If a married couple is buried side by side, one kit will suffice for both husband and wife. My family buys the pre-assembled bags and then supplements with an extra bunch of incense—enough to distribute three sticks per person among the family members gathered to pay respects.

All joss papers are to be incinerated into complete ash. This is where jumbo-size tin cans are extremely handy and sensible. To ease the burning of the joss papers, they can be folded into ingot shapes, which also enrich the departed.

To fold ingot shapes, take the square tan-colored joss paper and roll it into an empty tube similar to a coin wrapper. Tuck in the bottom edges and leave the top edges pointing out directionally, which creates the illusion of a boat-shaped gold ingot.

The red candles are on wooden sticks for pressing into the ground. But should the ground be too hard, use another large can filled with three to four inches of sand or some other granular item (my family uses kitty litter) for standing up the candles and incense sticks. Once the joss papers are burned and the candles lit, light the entire bunch of incense sticks at once. After they ignite, fan the flame out with your hand and let them smolder, because Auntie Lao frowns on blowing on lit incense with your breath. Distribute three lit sticks to each family member present. Stand the remaining incense sticks next to the red candles.

step four: giving respect

Now with three incense sticks in hand, family members form a single line. One by one, while holding the incense with both hands, each person pays respects by bowing three times to the grave and then places the incense sticks near the headstone, either in the ground or in the fireproof container.

The grave site ritual of laying out food and sending otherworld offerings is traditionally made to the direct line of ancestry, usually the grandparents and parents. Flowers are also laid for extended family members. In old China, paying respects was considered mandatory for agnate ancestors, or those belonging to the father's line of descent, and for a departed husband or wife. But my family doesn't distinguish between the two sides of kin and pays respects to both sides of the family. Because in America, we're separate but equal.

After all ancestral visits are made, everyone returns to the original site (usually the resting place of the parents or grandparents) and gathers up the canisters for Double Ninth Day (see chapter 7). My family leaves the food behind at the grave site, but it's also acceptable to repack the food for eating later. There is no set rule.

A Qing Ming Silent Prayer

Forefathers of our family,
of our kindred and of our race,
deign to accord us your protection.
All that we have is your gift,
all that we know is your knowledge
*   bequeathed—*
about laws of life and death,
about the things to be done and
the things to be avoided,
about ways of making life less painful
than nature willed it,
about right and wrong,
sorrow and happiness,
about the error of selfishness,
the wisdom of kindness,
about the need of sacrifice.
Unto you, the founders of our homes,
we humbly utter our thanks.

*   —Juliet Bredon and Igor Mitrophanow,*
*   The Moon Year*

> **A Note About Firecrackers**
>
> *Traditionally, firecrackers are lit during Qing Ming to frighten away evil spirits and bring about the sense of celebration, but the threat of fire has prohibited this practice in many American cities. Nevertheless, the Chinese find it hard to abstain from this custom, so don't be surprised to hear a few pau jeun in the Chinese cemeteries.*

otherworld money

Paper money is the otherworld's currency of choice. An old folk tale explains that otherworld currency originated from an old man-spirit who helped child prodigy Wan Bo win a literary competition. In the story, Wan Bo promised the old spirit whatever he wished if he helped him win first place in the contest. So the spirit asked Wan Bo to settle an old gambling debt with the Chang Lu spirit in exchange for the grand prize, and Wan Bo agreed. The young poet outperformed all others and won the competition, which amassed Wan Bo fame and national recognition. But in his glory, he forgot his promise to the old spirit. One day while Wan Bo was out on a walk, several black-winged creatures descended upon him, preventing him from moving a step farther. A flock of black crows had been sent to collect on the promise. As Wan Bo's memory was quickly restored, he fervently rushed down the path to the Chang Lu Temple to burn spirit money and joss paper to repay his debt and acknowledge his good fortune. Since then, joss paper and otherworld money have been considered the seeds for planting good fortune with the spirits.

yellow ribbons

Old Auntie Lao says that yellow ribbons keep away the wayward hungry ghosts who've been neglected by their wandering families. To shoo these spirits away so they won't skim off the top of a beloved's offerings, tie a yellow ribbon to the end of a bamboo stick and insert it into the earth at the grave site. Or simply secure the ribbon with a rock set atop the headstone. Auntie Lao says the ribbons will flutter like butterflies in the wind and assure the offerings will remain safe for the intended family spirits.

Here's more wisdom from Auntie Lao:

- *Six foods on a Qing Ming plate means the rains won't be late.*

- *Keep your wits and buy more bunches of incense sticks.*

- *Remember to make offerings to the Earth God so he won't steal from your family's heavenly stash.*

The Earth God resides in Chinese cemeteries to protect the house of the dead. Look for a simple, indiscreet stone on a well-traveled pathway near the center of the cemetery hill. Pay respects to this deity with a moment of recognition. To keep him satiated, leave flowers and morsels of food.

feast without fire

It was established that no fires should be lit on the eve of Qing Ming in honor of patriot Jie Zi Tui. This brave soldier and loyalist lived during the Warring States Period (475–221 B.C.E.) the same time as famous poet and statesman Qu Yuan (see chapter 3, "Dragon Boat Festival").

Jie Zi Tui, in a sacrificial act, saved his lord, the Duke of Wen of the Jin state, from starvation during a perilous journey when food supplies ran out. In desperation, Jie Zi Tui fed the flesh from his own leg to his ruler. The lord was so grateful that he promised to honor his rescuer handsomely upon returning home. But in the excitement and jubilation of returning safely, the duke forgot his promise to Jie. Angry and disappointed, Jie no longer could bear to remain in the Jin state. Late one night, without a word to anyone, he and his mother left town and fled to the mountainous forest of Mianshan.

When the Duke of Wen learned of Jie Zi Tui's disappearance, he remembered he hadn't fulfilled his promise to reward Jie. He asked Jie to return to service, but Jie refused. Insulted by Jie's obstinate behavior, the

Duke organized a search party, but it was unsuccessful in locating the soldier. Suspecting Jie might be in the mountains, the Jin army set fires in three directions, leaving one route open where he could escape. Jie would surely not let his mother perish. But the army miscalculated. When the curtain of smoke lifted, both son and mother were found clutching a willow tree with no life left in either of them. Grief stricken from the tragedy, the lord erected the first ancestral tablet ever raised and declared that on the eve of Qing Ming, the anniversary of Jie Zi Tui's death, no fires ever be set. He further proclaimed that eating cold food would warm all hearts and minds in remembrance of the lessons left behind by this man.

the power of the willow

Since the day when Jie Zi Tui was found clutching the willow, the tree has been considered to possess protective powers against evil spirits. It's often associated with new life and is emblematic of spring's new leaves borne of the sun's light. Because the willow grows heartily in nearly all climates, it's become a symbol of vitality that stands against approaching darkness.

The Chinese often use the willow as a charm for protection. The third emperor of the Tang dynasty (618–907 C.E.) established the practice of wearing willow wreaths to guard against scorpions. Others used the branches to ward off sickness because of the willow's connection to the sun, the Great Healer in the sky. In early times, women wore willow sprigs in their hair during Qing Ming to whisk away evil spirits. Today in America, this is still practiced by Chinese women and children during traditional funeral services of close family relatives.

The willow is also known to influence the good by drawing the ancestral spirits home and simultaneously repelling hungry ghosts. This belief explains why brooms were traditionally made of willow leaves. In the hot summer months, willow branches were also believed to bring rain. In old China, the villagers battled drought with a rain ritual called "catching rain" by placing willow branches in large earthen jars and writing rain prayers on slips of paper to be strung as banners across the village streets. It's said to have worked wonders.

qing ming checklist

Here's a list of everything you'll need for the ancestral cemetery rituals for the Chinese festivals of the dead:

Item	Explanation
Red Tapered Candles	A pair to signify togetherness that lights the way out of darkness.
Incense Sticks	It's believed the ancestral spirits travel through the thin smoky trails of incense. Widely used in all Chinese rituals to honor the dead, incense sticks are approximately 7 inches long. They look like sparklers and are sold in large bunches.
Otherworld Money	Paper currency of the otherworld. Known to be issued in huge denominations with the King of Hell as the governing treasurer.
Joss Paper	Square tissuelike tan-colored paper imprinted with a Chinese character. Joss paper is imprinted with silver and gold stamps because when gold and silver are compounded as Chinese characters they represent wealth

Item	Explanation *(continued)*

and riches. Joss papers can be folded into ingot shapes so the paper will ignite and burn more easily (see page 48).

Spring Flower Bouquets

One for each grave site visited as a sign of respect and remembrance. Flowers are typically laid for departed grandparents, parents, aunties, uncles, and even for single siblings. Those who are widowed care for their departed spouses. My family includes flowers for our "old aunties" who cared for us while growing up but were not related by blood. Leaving flowers for influential old friends or inspiring mentors is a generous gesture but not required. The traditional duty is to the family's elders.

Two Jumbo-Size Tin Cans

At the grave site, practicality suggests burning the joss papers and spirit money in a fireproof container. A second container filled with sand or another granular item (e.g., kitty litter) is useful for standing the candles and incense.

Matches or Lighters

A "must" for the occasion.

Water and Garden Shears

A gallon or two is very convenient, as are garden shears to trim the grass.

planning for qing ming

Begin planning for Qing Ming one to two weeks before this April festival. Some Chinese American families are flexible about the April 4–6 date and simply gather at the cemetery during Easter week out of convenience.

Activity	Suggested Timing
Notify family and friends of the date of remembrance.	1 to 2 weeks prior to gathering at the cemetery
Begin compiling your Qing Ming Checklist for gathering all needed grave site ritual items (for reference, see pages 55 and 56).	1 week prior
Purchase otherworld supplies (incense sticks, joss papers, red candles).	1 to 7 days prior
Collect containers for washing the grave sites and burning the ritualistic items.	1 to 7 days prior
Have a list and map ready for locating the burial sites of your loved ones.	1 to 2 days prior
Pack up the food, drink, flowers, supplies, and family for a morning of remembrance.	Morning of Qing Ming

righteous   *noon*   *festival*

As the heat rises

The waters flow

The heartbeat of the dragon calls

Qu Yuan, where have you gone?

Do not despair

We come in your honor

We come in your spirit

Qu Yuan, one of China's most impassioned poets from over two thousand years ago, lives on through the insistent beat of a dragon boat's drum, the synchronization of a crew, and the roar of a crowd during the Dragon Boat Festival. The fifth day of the fifth lunar month, also known as Double Fifth, is the second of three Chinese festivals that are widely celebrated and designated for the living. (The others are Chinese New Year and the Mid-Autumn "Moon" Festival.)

When China was primarily an agrarian state, Double Fifth was originally associated with the summer solstice. Farmers welcomed the warm season of monsoons by paying homage to the river dragon, the ruler of rainwater and streams, to assure a bountiful rice harvest in southern China.

double trouble

Double Fifth is regarded as the most dangerous day of the lunar calendar and is also considered an unlucky birthday. On the Gregorian calendar, the fifth lunar month typically falls around the latter part of May to mid-June. This fifth month traditionally was believed to be filled with disease and danger because the hot, humid weather brought with it many infectious diseases. Conditions became ripe for the Five Gods of Plague to cause havoc for humans as insects multiplied and the river water and streams grew stale from overuse.

During the fifth moon, yang energy is believed to be at its height, while the summer solstice initiates a transfer of yin forces. It's a time when nature stirs as the seasons play push-and-pull. Nevertheless, the annual balance of earth's energies creates an auspicious time for a festival filled with rituals to ward off bad luck and illness.

Opposites Attract 陰 陽

The balance of all things in nature comes from the Taoist concept of yin and yang taught by Lao-tzu, the Chinese philosopher who lived in the sixth century B.C.E. Yin and yang is a collective principle expressed through the harmony of opposites— masculine to feminine, sun to moon, hot to cold. Yin-yang is the natural order in which all things harmonically complement and balance each other. With yin, there is yang; with yang, there is yin. A woman is yin, but yang on the inside. A man is yang, but yin on the inside. Neither aspect exists without the other.

- Yin is the feminine phoenix, moon, water, passive, cool, mysterious, even numbers, north, and west.

- Yang is the masculine dragon, sun, fire, energetic, hot, direct, odd numbers, south, and east.

zapping evil

To counteract apparent threats of illness and misfortune, an ancient five-pronged strategy to deflect the tactics of the five evil gods was devised in old China. It included the following:

- Burning realgar to exterminate demons and bugs that multiply when the weather grows warm.
- Hanging a picture of Zhong Kui, the demon slayer, on the front door.
- Taping yellow strips of paper with exorcising messages onto the home's interior and exterior.
- Eating cake in the shapes of the five poisonous creatures (scorpion, snake, centipede, lizard, and toad) for protection.

- Displaying a bouquet of four kinds of leaves and one flower hung over the doorway to expel demons.

Old Auntie Lao's favorite old-world charms to avert evil are:

- Adding pointed leaves and a flower, such as the iris or the bird of paradise, to floral arrangements. The Chinese characters for pointed leaves translate to the words for slaying demons and are believed to remedy summer qualms.
- Hanging bunches of mugwort and other aromatic leaves on top of the front door with a piece of onion or garlic so the strong scents will drive away the demons.
- Hanging a mirror over the front door to reflect away any lurking evil.

a poet's life

The Dragon Boat Festival remembers Qu Yuan, poet and political loyalist of the Chu State who tragically perished through self-sacrifice during China's Warring States Period (475–221 B.C.E.). This was a brutal time of feudal lords involved in a maelstrom of plotting and scheming for political power. Alliances constantly shifted. Diplomacy was nonexistent.

Qu Yuan was the chief political counsel to the Chu ruler. Highly influential and intelligent, Qu was regarded as a political genius who always was consulted when the kingdom was confronted with issues of survival. However, despite Qu's political prowess, the Chu ruler did not heed his suspicions about Chu's neighboring rival: the Qin state. Qu continually advised the ruler against signing a treaty with the Qins but to no avail. Soon he fell out of favor.

Qu Yuan was subsequently banished to a desolate area of Hunan Province, where he became increasingly distraught and despondent. Despite his disgrace, he wrote poetry to express his unyielding love and loyalty for his kingdom and ruler. One day, on learning that his beloved kingdom had indeed fallen to its rival, the Qin state, he wrote the famous Chinese poem "Li Sao" ("On Encountering Sorrow") on the shores of the Miluo River, a tributary of the Yangzi. The loss was too great a burden to bear. With crumbled kingdom and broken heart, he anchored himself to a boulder and plunged into the rushing river.

As word of Qu Yuan's ultimate sacrifice spread throughout the village, fishermen raced into their boats in an attempt to save him. They desperately beat their drums and frantically paddled up and down the waters. To prevent the fish and the river dragon from devouring the poet, they furiously splashed the water with their paddles. But all efforts proved futile. In resignation, the villagers began the practice of tossing rice into the river to ensure that Qu's spirit would be well fed and that the fish would never feed on his body.

Qu Yuan's masterpiece, "Li Sao," is still widely treasured today. In 376 lines of verse, Qu Yuan reflects on the importance of family, politics, nature, morals, ideals, the need for leadership, opportunity, struggle, and the determination to be true to oneself despite all odds.

All Dragons Are Not Created Equal

_Dragons are believed to be mythical crea-
tures wielding great powers and are highly
respected by the Chinese. The energies of a
dragon can be tapped for a positive outcome that relates to the principles of yin
and yang and good and evil. These creatures
symbolize fertility, strength, and vitality,
with the most popular dragon types being
Long, Li, Jiao, and Mang._

_The Long dragon is the supreme fire-
breather and is the imperial symbol of the
emperor. This ruling-class dragon is recognized by its five claws and is believed to
possess immeasurable powers. While reclin-
ing, its head points south and its tail
stretches to the north. According to Taoist
legend, the Long dragon is a conglomeration
of nine common creatures: head of a camel,
horns of a deer, ears of a cow, neck of a
snake, stomach of a frog, scales of a fish,
eyes of a rabbit, claws of a hawk, and foot-
prints of a tiger. Long dragons lack the sense
of hearing, hence the deaf in southern China
are also referred to as long._

_Li is the water dragon that rules the
rivers, seas, rain, and directions of the compass. Offerings are made to this creature for
rain during the harvest season. This dragon_

*has no horns and breathes water and sea
foam.*

*Jiao is the earth dragon. In the concepts
of feng shui, it is believed that a land-
scape's hills and mountains are the Jiao
dragon's back and that by complying with
this dragon's energy, good fortune and success will be realized.*

*The Mang dragon was created by the im-
perial court to represent commoners. This
dragon is recognized by its three or four
claws and is considered ordinary.*

rice thieves

In 40 B.C.E., nearly two hundred years after Qu Yuan's death, it is said,
local fishermen received a visit by Qu's spirit, who reported that the rice
intended for him was being eaten by the fish and the powerful river
dragon. Qu advised the fishermen to wrap the rice in special packets tied
with threads in the imperial colors of red, blue, white, yellow, and black,
which corresponded to the five directions of south, east, west, center,
and north. The combination of the five colors served as an amulet that
was feared by the river dragon and would keep the rice safe for the spirit
intended.

Thus, the tradition of wrapping rice in bamboo leaves and tying the
packets with string was born. This Dragon Boat Festival culinary classic,
known as *joong,* is now available year-round in many Chinatown bakeries,
delis, and markets.

the art of joong (zongzi)

Today, the *joong*'s bamboo leaves are tied with white string to secure the fillings. Contents vary from sweet to savory varieties, and the only way to be sure the fillings are to your taste is to make them at home.

Annually in May to June, two to three generations of Chinese moms, aunties, and daughters come together to make *joong* as a way to acknowledge dragon boat folklore. It's also their way of passing the torch from one generation to another. *Joong* will keep extremely well for several months if frozen and can be shared with relatives and friends around the summer solstice.

The savory glutinous rice version is satisfying fare and can take the place of an entire meal. Common fillings include glutinous rice wrapped around salt pork or Chinese sausage, dried shrimp, salted egg yolk, peanuts, and chestnuts. A sweet version is an alkaline-preserved gelatinous rice paste that's the color of molasses. The center often holds sweet black bean paste. Those with a sweet tooth dip it in sugar or maple syrup.

In Southeast Asia, the Chinese *joong* is integrated with Malay influences to create a sweet, peppery tang. Once again, glutinous sticky rice is used. But the filling ingredients are finely diced lean pork, chestnuts, Chinese black mushrooms, winter melon, onions, scallions, and garlic, all bound together by a dark soy sauce that's liberally seasoned with sugar and pepper. It's a fine example of Nonya cuisine from the Straits Chinese who evolved from the intermarriage of Chinese immigrants and the Southeast Asian indigenous population.

Making *joong* in the summer is similar to Mexican households' tradition of making tamales at Christmastime. It's considered an enterprising activity. Preparing the contents requires some organization and preplanning. Folding the bamboo leaves around the fillings requires a quick lesson in the fine art of wrapping a three-dimensional triangle and obtaining the desired five points of direction—east, west, north, south, and center. Gather friends and family for an afternoon *joong* session. An assembly line always accelerates the preparation and cleanup.

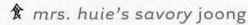 ## mrs. huie's savory joong

makes 40 packets

This hearty joong recipe is my personal favorite. It comes from my friend Jane Huie Lang's mother, Mrs. Susie Huie. Her secret is a Chinese-style pork marinade that permeates throughout the sweet rice. Although this recipe requires a four-day preparation, assembly, and cooking process, it's well worth the effort. Enjoy!

3 packages dried bamboo leaves
White kitchen string
½ cup Koon Chun potassium carbonate and sodium bicarbonate
  solution *(gon suey)*
8 pounds side cut of pork in ½-inch-thick slabs
½ cup salt
1 cup plus 1 tablespoon sugar
¾ cup *chee hou* sauce
¾ cup brown bean sauce with whole beans
½ cup oyster sauce
2 tablespoons dark soy sauce
¼ cup Shao Xing rice wine
1 pound whole raw chestnuts
1½ pounds skinless raw peanuts
40 salted eggs
3 pounds Chinese sausage
10 pounds Sho Chiku Bai sweet rice
⅔ cup canola oil
⅓ cup distilled white vinegar
1 pound medium or large dried shrimp
1 teaspoon white pepper
cheesecloth

Day One Preparation

Sort through the bamboo leaves by discarding the small leaves. Stack the leaves smooth side up and with the stems pointing in the same direction. Divide the leaves into three bundles and loosely tie them with white string. In a large pan, soak the bundles in cold water for 2 to 3 hours; turn the bundles occasionally in the water. Drain the water from the leaves. Add enough hot tap water to cover the bundles. Add the potassium carbonate and sodium bicarbonate solution to the hot water. To prevent the bundles from floating to the surface, weigh them down to keep them fully immersed in the water. Soak overnight.

Day Two Preparation

Drain the leaf bundles from the previous day's soaking. Add very hot tap water to the leaves and continue soaking for another 24 hours. Rub the sides of the pork slabs with ¼ cup of the salt and ½ cup of the sugar. Let the pork marinate in its juices for 6 hours. Drain the liquid from the pork. In a small bowl, combine the *chee hou* sauce, the brown bean sauce, ¼ cup of the oyster sauce, the soy sauce, the Shao Xing rice wine, and ¼ cup of the sugar. Add the Chinese marinade sauce to the pork and place the slabs on baking sheets. Cover the pork with cheesecloth and let cure at room temperature for 2 days. Rinse the chestnuts and peanuts in warm tap water and soak overnight in separate bowls.

Day Three Preparation

Drain the leaf bundles from the previous day's soaking. Add very hot tap water to the leaves and continue soaking for another 24 hours. Drain the chestnuts and remove any residual skin. Cover the chestnuts with plastic wrap and refrigerate. Drain the peanuts, cover with plastic wrap, and refrigerate. Separate the salted egg yolks from the egg whites by cracking the eggs one at a time into a bowl. Pick out the yolks and discard the whites. Cover the yolks with plastic wrap and refrigerate for the next day's assembly. Cut the Chinese sausage on the diagonal into 2½- to 3-inch-long pieces. Cover the Chinese sausage in plastic wrap and refrigerate. Wash the rice in water until the water runs clear. Soak the rice in cold water for 1 hour. Add the chestnuts and peanuts to the soaking rice, toss until evenly distributed, and soak for 1 hour more. Drain the rice, chestnuts, and peanuts and let the mixture sit in a colander overnight.

Day Four Assembly and Cooking

1. Transfer the rice mixture into a large pan. Combine the remaining ¼ cup salt, 3 tablespoons of the sugar, ⅓ cup of the canola oil, and the vinegar. Thoroughly mix the seasoning into the rice mixture. Set the rice mixture aside for assembly. Using a frying pan or wok, heat the remaining ⅓ cup canola oil and sauté the dried shrimp. Add the remaining ¼ cup oyster sauce, the remaining 2 tablespoons sugar, and the white pepper. Let cool and set aside for assembly. Cut the cured pork into 1-by 2-inch slices. Set aside for assembly.

2. To assemble the *joong* packets, take 1 leaf and place vertically, smooth side up, with the stem at the top, on a clean work surface. Take a second leaf and place, smooth side up, with the stem at the bottom, on top of the first leaf slightly off-center by covering three-quarters of the first leaf and providing 1-inch additional leaf coverage on the right side. Fold both leaves in half with the smooth side facing in. With the fold at the bottom, fold the bottom left corner of the leaves under by approximately 1 inch, forming a pocket from the right-side opening. With your left thumb and index finger, hold the bottom left corner securely and spoon ¼ cup rice mixture into the pocket. Take a diagonal cut of the Chinese sausage and bury it into the rice toward the corner. Add 5 dried shrimp, 1 salted egg yolk, and 1 slice of pork. Wrap 2 more leaves around the

package in slightly varying positions to provide enough leaf coverage for securing the contents, while still maintaining the opening. Add ½ cup rice mixture to cover all of the other ingredients. Wrap another leaf around the packet to seal off the edge. Close by folding down the open end of the leaves. Secure the packet by winding the string three times vertically around the *joong* from one end to the other, turn horizontally to wind the string around two to three times, and double-knot. Repeat with the remaining bamboo leaves and filling ingredients.

3. Stand the *joong* vertically at the base of a large pot. The *joong* can be stacked as long as there is enough height for boiling. Add warm water to the pot with a 2-inch coverage over the *joong*. Bring to a boil over high heat. Reduce the heat to medium-low and let cook for 3 hours. Skim the surface foam as needed. The *joong* must be fully immersed in water during the cooking process; add boiling water if necessary. When the packets are plump and the rice is soft, remove the *joong* from the water. Drain the *joong* completely by turning on both sides. Rinse the packets under warm water to remove any excess foam and oil from the cooking process.

4. Serve the *joong* hot by carefully removing the string and leaves. *Joong* can be refrigerated for up to 1 week and they freeze extremely well for several months. To reheat, boil in water for 20 to 30 minutes until hot.

dragon boat fever— it's contagious

Today, the dragon boat celebration has evolved into a sporting season that spans beyond the fifth lunar month. Dragon boat festivals are held from February to October in many corners of the world, including both ·

coasts of the United States and the vast international waterways of Canada, Europe, and Asia. The main attraction is the dragon boat races that have expanded from a regional event into an international racing competition that is favorably being considered as a demonstration sport for the 2008 Beijing Olympic Games.

Dragon boat racing is one of the earliest forms of boat racing. The tradition originated over two thousand years ago when dragon boats raced out into the river in an attempt to save the poet Qu Yuan. Now dragon boat racing continues to grow in popularity as an accessible, invigorating water sport for all ages and divisions of competition. Whether you are a paddler or a spectator, the races will capture your breath and spirit.

The most prestigious dragon boat competitions include the World Dragon Boat Championships, an international competition featuring national crews that is held in odd-numbered years, and the World Club Crew Championships, which challenge the world's best clubs in even-numbered years. The International Dragon Boat Federation, the governing body for the sport, is the sponsor of both championships. Venues for these worldwide competitions also alternate and have been held in world metropolitan centers such as Hong Kong, Nottingham, Auckland, Vancouver, Rome, Cape Town, and Yueyang, China, considered the birthplace of dragon boat racing.

It's estimated that in the United States, over forty festivals are held in cities that include New York, Boston, Houston, Los Angeles, Miami, Phoenix, Portland, and San Francisco. The two largest festivals in North America are held in Toronto and Vancouver. (See "Dragon Boat Race Venues and Associations," page 74.)

Many of the largest races begin with a blessing of the boats by priests who conduct a Taoist ceremony that incorporates prayers, incense, and firecrackers. In assembling the boat, the ritual of painting the dragon's eyes gives sight to the vessel. Connecting the head at the bow and the tail at the stern readies the boat for competition. But the element that brings the vessel to life is the drum that symbolizes the beat of the dragon's heart.

Dragon boat teams paddle with the same fervor as the fishermen who fought to rescue Qu Yuan. Categories of teams range from novice, recreational, and competitive to men, women, mixed, seniors, and youth. Teams compete in fifty-foot-long teak or fiberglass dragon boats that typically carry a crew of twenty paddlers, plus a drummer and a steersperson who sends commands to hold steady or surge forward. The drummer, the heart of the dragon, maintains the team's rhythmic pace and provides a source of inspiration when the pain and exhaustion begins to set in. The two strokes in the front row set the cadence to orchestrate the team, while the "engine," the true power of the boat, represents the team's biggest and strongest paddlers. The course spans 500 to 1,300 meters, but unfamiliar waters and unfavorable weather make the course unpredictable.

dragon boat race venues and associations

Dragon boat racing weaves elements of legend, culture, and competition for a seasonal experience that fuels the spirit with lots of splashing and socializing. As one of the fastest-growing sports, dragon boat associations and teams are working together toward worldwide recognition and local participation. Regional races are held in all waters ideal for a challenging course, such as the Mississippi River, lakes, estuaries, channels, and bays. Check your local newspaper and the Web for dragon boat festival and race sites near you.

Here are some of the largest annually held dragon boat competitions and spectator events in North America:

- Toronto International Dragon Boat Race Festival—considered one the largest dragon boat festivals in North America, attracting nearly two hundred teams and over 250,000 spectators during its June race weekend.
- Alcan Dragon Boat Festival—Vancouver dragon boat competition held in June that is similar in scope and size to Toronto's event.
- New York City International Dragon Boat Race Festival—late-July weekend event with over one thousand paddlers and tens of thousands of spectators.
- Portland-Kaohsiung Sister City Association Dragon Boat Races—held in June in conjunction with Portland's Rose Festival and is attended by over one hundred international and national crews.
- Northern California International Dragon Boat Championships—held in San Francisco in September with nearly one hundred national and international teams in competition.

For further information on the sport of dragon boat racing and the qualifying competitions, contact the dragon boat associations below:

- International Dragon Boat Federation at www.dragonboat.org.uk
- United States Dragon Boat Federation at www.usdbf.com
- Pacific Dragon Boat Association—West Coast Region at www.pdbausa.org
- American Dragon Boat Association—Midwest Region at www.americandragonboat.com
- Eastern Regional Dragon Boat Federation—East Coast Region at www.erdba.org

- California Dragon Boat Association at www.cdba.org
- Dragon Boat Canada at www.dragonboat.ca
- Hong Kong Dragon Boat Association at www.hkdba.com
- Asian Dragon Boat Federation at www.asiandbf.org
- European Dragon Boat Federation at www.dragonboat.org.uk

chapter four / double seventh day

seven   month   first   seven

Maidens,

When seeking beauty and releasing

wishes for a kind husband, remember

that your true beauty lies within.

Double Seventh Day draws on the romantic Chinese folktale of the Cowherd and the Weaving Maiden. On Seven-Seven, these clandestine lovers, eternally separated from each other except on this one night, are reunited by a bridge of magpies. Although the festival of Seven-Seven is not widely observed in America, it's a holiday full of ritual to satisfy the heart's romantic desires.

the cowherd and the weaving maiden

According to legend, the Weaving Maiden was the seventh daughter of the Sun God of Heaven. Extremely talented in needle crafts, she was deemed the master seamstress and weaver for the heavenly gods. One day after laboring for hours, the young maiden descended to earth with her six older sisters to bathe in a stream. As the maidens relaxed in the pristine water, a young Cowherd in the nearby meadow saw the maidens from a distance and stood enchanted by their beauty.

As the Cowherd passed the stream, an immortal disguised as a cow instructed the young man, "Master, the seventh is beautiful, talented, and kind-hearted. She spins heavenly silk for the gods and presides over the weaving of earth's maidens. Quickly, go take her heavenly robe and make her acquaintance. Should you become her husband, you will gain immortality."

The Cowherd saw the clothes that lay nearby on the grassy knoll and hurriedly gathered the youngest maiden's elegant silk robe. When the Weaving Maiden emerged from the stream to collect her garment, she discovered it gone. As she looked about, the Cowherd approached with her robe gently draped over his arm. Relieved, she quickly took her silk covering and threw it over her shoulders. When her grateful eyes met his, her heart told her she'd just met her soul mate, and she decided to remain on earth with him.

The couple married and lived happily for many years, but with the passing of each year, the heavenly deities grew anxious. So in love and immersed in each other, the lovers began to neglect their talents and responsibilities. The Weaving Maiden ceased her needlework, and the Cowherd ignored his livestock.

Angry at all the negligence, the Heavenly Mother demanded that the Weaving Maiden return home. As a daughter of the heavens, she obeyed, sacrificing her own happiness. The Weaving Maiden retrieved her heavenly silk garments and ascended to the sky. The Cowherd attempted to follow by flying up into the heavens, but the Heavenly Mother barred his way. She waved her powerful silver hairpin to create a silver river of stars so wide that he was unable to cross.

When the Sun God heard of his youngest daughter's forced separation, he was saddened immensely. Out of compassion, he allowed the couple to see each other one night each year on the seventh moon, seventh night. If the skies were clear, the magpies of heaven would form a bridge over the silver river—known as the Milky Way—so that the Cowherd and the Weaving Maiden could reunite.

With the exception of this single night, the lovers forever live in separate constellations that can be seen in the sky. In the universe, the Cowherd is the star Altair in the Aquila constellation, while the Weaving Maiden is Vega in the Lyra constellation—two stars separated by the Milky Way.

clear skies

On Seven-Seven, villagers prayed for clear skies so that the magpies could reach the heavens to reunite the Cowherd and the Weaving Maiden by forming a bridge across the Milky Way. Should it rain, it's said, the magpies would be prevented from gathering and the couple's tears would fall as rain until they could meet again the following year.

patron saints

While the Cowherd is considered the patron saint of cattle and livestock, the Weaving Maiden is recognized as the saint of the home and needlework. Chinese women pray to her for a happy marriage and the gift of many sons. It's also believed the Weaving Maiden is the protector of orphaned girls and possesses great compassion for the plight of all young women.

a maiden's ritual

On the evening of Seven-Seven, maidens traditionally gathered to invite good blessings in marriage and family. Based on old folklore, women would honor the seven sisters of heaven, daughters of the Sun God, as a way to bring a good spouse, a happy marriage, and the gift of many sons. It was an evening when fortunes were sought and omens were read to foretell beauty and skill.

Maidens prepared family altars with the finest embroidery and needlework. The altar included seven plates of food, one for each heavenly muse. Typical offerings included fruit, young bean or rice shoots, sweets such as candies and cookies, cups of tea, flowers, sewing items, and hair and beauty products. Ideally, groups of seven girls would gather to receive the blessings from the celestial Seven Sisters, and in return, powers of irresistibility and attraction would be granted.

the sewing season

In old China, the seventh month was associated with a change of seasons. With the transformation of colors and falling leaves, autumn became a reminder of nature's impermanence. In anticipation of winter, warm clothing for the family was sewn and preparations were made for the harsh cold. Fabric, needle, and thread were the instruments of necessity for the women of old China. Sewing ability was highly revered, and a woman's worth was often determined by how adept she was at needlework. How were a young maiden's sewing talents gauged? Very long ago, two ways to foretell a maiden's proficiency were by reading the shadows of floating needles and by using spider boxes.

a needle's shadow

It's said that analyzing the shadow a needle casts in a bowl of water offers insight about a maiden's expertise in needlework. If the shadow looks like a leaf or flower, the maiden will be highly proficient. But if the shadow resembles a thin stick, she should seek work in the kitchen instead.

spider boxes

While shopping for Chinese souvenirs, small wooden boxes can be found with a dangling plastic black spider as their primary resident. These boxes come from the folk tradition of determining a maiden's sewing expertise through the web a spider spins in a small box on the eve of Seven-Seven. On the morning of the festival, each maiden would open the box to see if her spider had been productive during the night. If a web had taken form, it was an omen for great success as a master seamstress.

deliverance   *orchid*   *festival*

(of Buddhism)

Always hungry, always needing, she would have to beg food from other ghosts, snatch and steal it from those whose living descendants give them gifts. She would have to fight the ghosts massed at crossroads for the buns a few thoughtful citizens leave to decoy her away from village and home so that the ancestral spirits could feast unharassed. . . . My aunt remains forever hungry. Goods are not distributed evenly among the dead.

—Maxine Hong Kingston,
The Woman Warrior

According to the Chinese almanac, the gates of the underworld are opened on the seventh month, releasing the hungry ghosts to roam the earth on holiday. To pacify these forgotten souls, the Chinese provide offerings to them on the fifteenth day (Seven-Fifteen), during the Festival of Hungry Ghosts. It's a Chinese version of All Souls' Day, when the living appease the ghosts with a feast all their own.

Hungry ghosts live in a continual land of limbo. They belong neither to the living nor to the dead. These ghosts are caught within the courts of justice and don't hold the full-fledged status of spirits of the otherworld. They are often neglected, having been victims of unfortunate or premature death.

To placate these roaming ghosts and prevent them from wreaking havoc, makeshift altars filled with assorted offerings are set up in public places—along roadsides, at intersections, and in temples—but never close to home! Giant joss sticks glow at open food-market stalls, and card tables display offerings of cold uncooked foods such as fruit, rice, and raw meat. Joss paper representing clothing, shelter, and spirit money is burned to deliver basic supplies. Lit incense smolders through the night. All hungry ghost offerings are relatively simple and not as elaborate as those sent to the family's direct ancestors.

floating lanterns

In prerevolutionary China, it was ill-advised to disregard the day of the dead. All boats were anchored and no fish were caught. On Seven-Fifteen, lanterns in the shapes of lotus flowers and small glowing boats would be released on the waterways at dusk. This widespread custom is in remembrance of those who perished by drowning and to give light to all spirits in the underworld. Peasants would assemble on the shores and release into the darkness miniature boats containing flickering candles so the waterways could shine back to the sky's constellations. It was a silent poem for the eyes.

Once all the floating lights were extinguished, the villagers would return home in the hope that a wandering spirit had caught the brightness to journey into another realm.

> **Yulan *Temple Ceremony***
>
> On Seven-Fifteen, Taoist and Buddhist priests and nuns perform the Yulan ceremony with chanting and prayer. The Chinese dutifully gather at temple out of compassion for all dead souls, whether hungry or well nourished, to remember the spirit world by lighting incense and burning joss paper and otherworld money.

mu lian and the hungry ghosts: a buddhist tale

Mu Lian was a Buddhist disciple of Sakyamuni. Like the Buddha, Mu Lian was also a royal son who forsook inherited wealth to become a monk. But he was a compassionate and gifted being who possessed magical powers. Devoted to his gravely ill mother, Qingti, Mu Lian fed her soup containing meat to restore her health. But, upon discovering she had broken her vows to vegetarianism, she was so overcome with shame that she immediately committed herself to the underworld. Mu Lian attempted to find and care for his mother's spirit but discovered she had become imprisoned in the land of hungry ghosts. Whenever he sent her offerings, the other ghosts devoured them. When morsels finally reached her, they erupted into flames. Disheartened by guilt about his mother's predicament, Mu Lian sought to punish himself to death in order to free her spirit. But when the Buddha learned of his disciple's desperation, he provided Mu Lian with another way.

The Buddha gave Mu Lian two items: a golden staff for opening the gates of the underworld and instructions for traveling through its murky maze. As Mu Lian successfully navigated through the gates of fire, he encountered evil wardens and demons of the underworld. Although he finally located his mother, he was still unable to free her from her suffering. So Mu Lian again sought counsel from the Buddha, who this time gave Mu Lian the *Yulan* prayers written to save starving spirits and allow them new life. Subsequently, Qingti was released from the underworld to repent her wrongs and move up the path to reincarnation. At last, Qingti attained a place of peace.

Thereafter, Mu Lian dedicated his life to offering special prayers and food for all trapped hungry ghosts. He believed that by lessening their misery, he could help them ultimately achieve fulfillment and rebirth.

di zang pusa

It's said that the sage Mu Lian was an incarnation of Di Zang Pusa, the redeemer who descends to the underworld to release lost souls. This lord's image is represented with eighteen hands, embracing arms of mercy, and gifts of the Buddha: a staff to open the gates of the underworld and a jewel to shed light on all anguished souls.

The lunar date of Seven-Thirty, or seventh month, thirtieth day, is considered Di Zang's birthday and marks the end of the seventh moon. It's believed that Di Zang was a powerful deity and protector to

all who suffer. Chinese Buddhists believe that within every human being resides the key to inner nature (*di*), the jewel of pure thought
(*zang*), and the staff of strong character. Then it is understandable that
the image of Di Zang is rarely depicted, as his spirit lives unseen
within the heart.

mid   *autumn*   *festival*

To me, the Moon Festival means a round jewel of pastry the color of molasses. At the center is an orange yolk. On the full moon's brilliance, I release all my secret wishes to Chang-E. I think of life's orbit. Sun and moon. Past and present. With the moon's reflection, I'm reminded of who I am, from where I come, to where I go. And I give thanks.

The Festival of the Mid-Autumn Moon is celebrated on the eighth lunar month, fifteenth day (Eight-Fifteen), which typically falls in September on the solar calendar. It's a Chinese holiday of thanksgiving when wishes are sent to the Moon Goddess, moon cakes are shared after a family dinner, and the year's brightest moon is worshipped.

In old rural China, the Moon Festival was an occasion for celebrating a bountiful harvest. Villagers were spiritually attuned to the lunar calendar for planting and harvesting the land's crops. It was a time when farmers finished laboring in the fields and families reunited to give thanks before the chill of winter. The arrival of the autumn moon gave way to the yin (feminine) principle to influence nature as the days grew cool, dark, and wet.

The moon is at its brightest and fullest of the year on Eight-Fifteen, when the earth and the moon are in closest proximity. The moon is the soul of the festival. Its full roundness symbolizes the unity of a family circle, the cycle of life, and the old folklore that connects the past, present, and hereafter.

chang-e, the moon goddess

The moon's realm belongs to the Moon Goddess, Chang-E. Known as Chang-O or Shiang-O in Cantonese, she was a beautiful servant girl of the Jade Emperor in the Heavenly Palace. But her beauty did not serve her well, for when she broke the emperor's beloved porcelain jar, she was banished to earth to be a daughter of a poor farming family. She eventually married a young hunter, an expert archer named Hou Yi.

One day, ten suns rose together in the sky. Their pulsing heat threatened to parch and destroy earth. Being an expert marksman, Hou Yi was recruited by the Heavenly Gods to rescue the planet. He successfully shot down nine suns, leaving one for earth's nourishment and warmth.

As a reward, the Jade Emperor named Hou Yi "the Divine Archer, Ruler of the Solar System" and gave him the elixir of immortality. Hou Yi promptly hid the elixir and immediately began his rule. Unfortunately, fame filled Hou Yi with hubris and he soon became an oppressive, arrogant tyrant. Nothing Chang-E tried changed her husband's behavior until she discovered and drank the concealed elixir glowing on a shelf. In the moment she swallowed it, Chang-E grew weightless and began to rise toward the palace window. Hou Yi desperately tried to reach for her, but

Chang-E continued to float toward the moon and beyond Hou Yi's grasp. Helpless and distraught, he watched his wife drift away. But as she rose higher and higher, an unbelievable sight occurred before his eyes. Chang-E began to transform into a frog, the yin creature of the lotus pond.

When Chang-E landed on the moon, she grew ill and coughed up the luminescent elixir. The magic elixir instantly transformed into a rabbit as jade-white as the moon. This was the birth of the White Rabbit, also known as the Moon Hare or Jade Rabbit, Chang-E's constant companion on the moon. To this day, it's said, the White Rabbit dwells in a forest of cassia trees, pounding cloves and cinnamon with a mortar and pestle in the vain attempt to re-create the elixir of immortality.

Resolved to accept his wife's newfound status as the Moon Goddess, Hou Yi had the Moon Palace built for her among a grove of cinnamon trees. The memorial shimmers with brilliance in the night sky. Now and forever, the heavens permit the Divine Archer and Chang-E to reunite there once a month on the full moon, the fifteenth day of each lunar month.

The toad and the moon are considered yin symbols relating to water.

Wh-abbits

Based on early Buddhist teachings, the white rabbit is revered as a sensitive creature with the gift of healing. It's said that when all the forest creatures prepared to greet the Buddha, they gathered and hunted for great offerings to the Enlightened One. However, seeing the entire collection of gifts presented by the other creatures, the rabbit was ashamed of his meager offering of grasses and herbs. Feeling that there was no other recourse, he made the ultimate offering by leaping into the fire in front of the Buddha. There's no question that he made an unforgettable first impression.

Moon's Woodcutter

Wu Gang, the woodcutter, is also a resident of the moon according to Chinese mythology. His sole purpose is to chop down the moon's Tree of Immortality, the cassia. But each time he strikes the tree with his ax, the wounded tree magically heals. This perpetual act is a life sentence the Jade Emperor imposed on Wu Gang for selfishly seeking immortality.

moon cakes

Moon cakes symbolize the heavenly blessings of longevity and good health. During the Moon Festival, round moon cakes are imprinted with floral designs or classic folktale figures, such as the Moon Goddess and the Jade Rabbit, from special wooden molds. Known as *yuet beng* in Cantonese, moon cakes are ideal gifts for family and friends. The Chinese in-laws would be impressed. By late August, cakes in decorative boxes—containing a few to many—dominate the shelves of Chinese bakeries and Asian food stores. Varieties are just as numerous: sweet bean paste, lotus seed paste, sugared melon, dried fruit, and seed and nut. The center can be filled with a whole duck egg yolk in single, double, and even triple versions, which are considered extra lucky. The yolk symbolizes the moon, and the egg's saltiness intermingles nicely with the flavor of sweet bean paste.

In addition to moon cakes, thick wheat-flour cookies shaped as fish, crabs, the God of Longevity, dragons, and pigs are available for children, who find them as hearty as teething biscuits. Especially look for pig cookies that are enclosed in brightly colored plastic baskets with red carrying string.

celebrating the harvest moon

The Empress Dowager Ci Xi of the Qing dynasty (1644–1911 C.E.) established the thirteenth to the seventeenth days of the eighth month as the official time to celebrate and fully appreciate the mid-autumn harvest moon. Most families today select one night during the full-moon period to dine and to moon-gaze.

A well-rounded evening of feasting consists of a family dinner followed by dessert taken when the moon is at its height of intensity. Typically, dinner of the family's favorite dishes is served banquet-style. Dishes can include a few appetizers, such as soup and a cold meat platter, and five to seven symbolically significant ones such as whole chicken, duck, crispy roast pork, beef, fish, seafood, and a vegetable. Rice is a requisite. The total number of dishes should total five, seven, or nine, which are yang numbers and considered lucky.

Below are some traditional foods eaten during the Mid-Autumn Festival:

- Taro—A starchy root vegetable that was first discovered by Ming dynasty (1368–1644 C.E.) soldiers in the bright moonlight of the fifteenth day, eighth lunar month. This potato-like vegetable saved the soldiers from starvation when defending China's coastline.

- Pomelo—Also known as Chinese grapefruit. This lemon-colored fruit is usually more pear-shaped, but once the skin and pith are peeled away, a round fruit with citrus segments remains. The sweet fruit is believed to ward off evil and promote good health. The Qing dynasty proclaimed the leaves to be sacred. Pomelos are a popular mainstay for nearly all Chinese festivals.

- Snails—Edible snails are eaten to celebrate a 1700s truce between two battling neighbors in southern China's Guangdong Province. It's said that two farmers were plagued with uncontrollable garden pests—snails. Each farmer thought he had found the ideal solution: to dump the snails into the other's fields. But soon, the bickering turned to blame, and serious fighting ensued. It took a wise Qing dynasty magistrate to resolve the dispute by cooking and serving the mollusks to the unsuspecting squabblers on the fifteenth day, eighth month. They were delighted with the tasty morsels, and soon the snails became a valuable commodity during the Moon Festival. Today, snails are eaten as a reminder of earth's wealth.

食 sautéed snails with black bean sauce

serves 6

This is very traditional rural southern China fare. Edible snails can be purchased in Chinatown fish stores. They are often stored in plastic buckets. For an authentic cultural experience, eat these delicacies by placing the open end of the shell to your mouth and sucking out the tender meat. It's well worth every effort to cook this dish because snails are a delicious treat, although they may be considered exotic by American standards.

> **1 pound live cultivated snails**
> **½ cup cornmeal**
> **2 tablespoons vegetable oil**
> **3 cloves garlic, minced**
> **2 tablespoons salted black beans (doe see), washed and mashed**
> **1 tablespoon hot chili sauce**
> **½ teaspoon salt**
> **½ cup water**
> **1 tablespoon cornstarch**
> **2 tablespoons chopped fresh cilantro**

To rid the snails of grit and sand, cover them with water and mix in cornmeal. Soak at least 1 hour, changing the water once or twice. Rinse after soaking. Use a pair of pliers to break off the snails' operculum—a hard disk covering their flesh at the end of the shell. Rinse the snails well under cold water. To a heated skillet, add the oil, garlic, black beans, chili sauce, and salt. Sauté for 2 to 3 minutes. Add the snails and cover and cook for 10 minutes over medium to low heat. Mix the water and cornstarch together and add to the sauté pan to thicken the juices. Garnish with the fresh cilantro. Serve at the family feast.

dessert by moonlight

The time for paying homage is after dinner. A table is placed out-doors—or by a window in cold weather—with an offering of thirteen moon cakes stacked in a pyramid. Because a complete lunar year contains thirteen months, the stack signifies happiness for the entire year. The moon cake's roundness symbolizes a complete family circle. Ancestors are remembered by burning incense, lighting candles, and bowing three times at the family altar. Here are some additional symbolic items placed on the Moon Festival table:

- Gourds for long-lasting togetherness
- Apples for peace ("apple" sounds like the word "peace" in Chinese)
- Pomegranates (for their seeds; mean many children)
- Asian pears, persimmons, grapes, peaches, melons, and other round moon-shaped fruits
- Soybean plants for representing the heavenly cassia tree
- Peanuts for long life ("peanut" sounds like the word for growth); also for the observation that a peanut is made when a peanut blossom buries itself into the ground and a new root sprouts
- Coconuts for promoting a healthy face and figure
- Watermelon seeds for many children
- Tea service complete with tea pot and cups

At the height of the full moon, the family dines on snacks, fruit, tea, and moon cakes. Paper lanterns in the shapes of rabbits, fish, birds, butterflies, and horses may decorate the home and please the crowd. A horse lantern is especially meaningful, as it's believed that the moon moves with the speed of a horse. Today, lanterns are also made in shapes of airplanes, rockets, tanks, and popular cartoon characters. The evening closes when everyone has taken time to release secret wishes to Chang-E.

> *A secret wish is your heart's deepest desire that is left unsaid.*

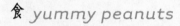 ## yummy peanuts

serves 4 to 6

This is a traditional and delicious way to prepare and season peanuts. The slow baking process lengthens their shelf life so that they can be eaten plain or used as an ingredient in other dishes. A symbol of a good and long life, peanuts are the perfect accompaniment to sweet, rich moon cakes and hot tea.

¾ **cup sugar**
8 **whole star anise**
6 **tablespoons salt**
4 **cups water**
2 **pounds raw skinless peanuts**

Combine the sugar, anise, salt, and water in a bowl. Add the peanuts and soak overnight. Drain in a colander for 6 to 8 hours. Preheat the oven to 250°F. On a cookie sheet, lay the peanuts in a single layer. Bake for 1 hour. Reduce the heat to 200°F and continue baking for 2 to 4 hours. After 2½ hours, test for doneness by letting some peanuts cool for 5 minutes. If crisp, remove from the oven and let cool completely. Store in an airtight container.

Moon Chat with Auntie Lao

Auntie Lao celebrates the full harvest moon with tea, moon cakes, sweets, and thanks. Here are some of the old village practices and beliefs she still lives by today:

- *Giving moon cakes for a season of joy and hope. Include egg yolks to symbolize the moon and family unity.*
- *Sharing moon cakes by cutting them into quarters before serving.*
- *Remembering the old Moon Minister of Marriage for an auspicious heavenly match (see page 29).*
- *Keeping an ivory complexion by placing face powder on the family's altar.*
- *Believing every watermelon seed eaten is a penny earned.*

moon cakes to the rescue

Moon cakes are recognized and honored for the role they played in Chinese history. Instrumental in the overthrow of a cruel Mongol state, Han leader Liu Fu Tong used moon cakes to distribute secret messages to his allies. The successful scheme ultimately led to an organized rebellion against the ruling Yuan dynasty (1279–1368 C.E.), the descendants of Genghis Khan.

On the night of Eight-Fifteen, Liu Fu Tong's rebels arranged for every baker and cake peddler to insert into thousands of moon cakes a small piece of paper with the date and time to rise for battle. The message instructed every family to raise a triangular flag with the seven stars of the Big Dipper and place lit lanterns on their rooftops at night between eleven and one o'clock. Villagers were also to beat on drums and gongs. When the Mongol soldiers saw the numerous flags and lights and heard the aggressive drumming, they were fooled into retreating for fear of being outnumbered.

The Han leader's clever efforts resulted in the successful revolution that allowed the Ming dynasty to commence its reign in 1368. Today, lanterns are lit in homes and hung out on the night of the full moon to celebrate freedom, peace, and unity.

Local Autumn Moon Celebrations

Many Chinese communities and groups in Chinatown celebrate the Mid-Autumn Festival with street fairs and parades. Often included in the festivities are activities such as moon cake tastings, ribbon dances, martial art demonstrations, and Chinese arts and crafts. Check local media listings in early September for Moon Festival celebrations near you.

moon festival planning

Preparation for the Mid-Autumn Festival uses many of the same organizational and planning skills as hosting a celebratory dinner party. Some families eat a banquet-style dinner in a restaurant and then return home for moon cakes and tea. Here's a checklist to get your full-moon celebration started:

Activity	Suggested Timing
Purchase or bake moon cakes for giving to close friends and relatives. Moon cakes often come in boxes of four and are shared by cutting each into quarters.	1 to 2 weeks prior to the Eight–Fifteen holiday
Select paper lanterns for displaying. Purchase ritual offering items (such as incense and candles).	1 to 2 weeks prior
Invite the relatives for dinner.	1 to 2 weeks prior
Plan and shop for the dinner menu and the Moon Festival table, which typically includes moon cakes, fruit, sweets, and tea.	2 to 5 days prior
Prepare the family's Moon Festival dinner.	2 days prior until Moon Festival day
Set the Moon Festival table and hang the paper lanterns.	Moon Festival day
Celebrate the fullest moon of the year by dining with family, making wishes, and giving thanks.	Moon Festival evening

double   *sun*   *festival*

重陽節

Ask me about Chong Yang, and I see chrysanthemums of bumblebee yellow and gladiolas of candy-apple red. I feel fall's briskness whisking through the hillside's eucalyptus trees. Three bows of respect to ease the cold of my ancestors. A watchful grandmother with a jade broach. A generous grand-father with a compass in his pocket. The little brother we never knew. All gone. Ever present.

Double Ninth Day is the ninth lunar month, ninth day, and is known as Chong Yang, which means "Day of the Double Sun." It represents the end of autumn, a season the lunar calendar places in the seventh, eighth, and ninth months. Like many Chinese holidays practiced to beckon good fortune, Double Ninth focuses on advancing success, inviting long life, and escaping danger.

Chong Yang is most commonly observed by Chinese Americans as a second Memorial Day of the year, when the family visits the grave sites to ready the spirits for winter. On this remembrance day, the traditional rituals of Qing Ming, the Clear Brightness Festival (see chapter 2), are repeated. Food offerings include an array of favorites such as steamed pork buns, golden sponge cake, crispy-skinned roast pork, whole boiled chicken, oranges, custard tarts, and tea. Chinese wine or liquor is also poured. Freshly cut flowers are laid at the headstone, and otherworld joss paper, money, candles, and incense are burned. Each family member takes three sticks of lit incense, gives three bows of respect, stands the incense at the headstone, and then moves to the next grave site. This ancestor-honoring ritual is usually practiced for departed grandparents and parents, but flowers are often left for aunts and uncles, too.

fall flower

The majority of fall floral bouquets that blanket the grave sites include the flower of the autumn moon—the chrysanthemum—often seen in blinding yellow. Considered with the plum blossom, the orchid, and bamboo to be one of the "four gentlemen" of the garden, the chrysanthemum has long been a symbol of longevity and good health to the Chinese and is a popular motif used in many Asian handicrafts and art designs.

> **When the chrysanthemum is gone, there are no others left in the world.**
>
> **—Chinese proverb**

> *Some common Chong Yang customs to invite success are drinking chrysanthemum wine, eating cake, hiking, and enjoying poetry.*

the great leap

Chong Yang originated during the Han dynasty (206 B.C.E.–220 C.E.) with the saying *deng gao*, which literally means "ascending heights." One day after a Han scholar, Huan Jing, was warned of an impending personal disaster, rather than alarm his family, he took them for an afternoon in the hills to picnic on cakes and read poetry. When the family returned from their outing, they found their home plundered and the chickens and livestock slaughtered. That the family escaped harm was due to Huan Jing's keen foresight and nimble action.

Through the centuries, *deng gao* has become a metaphor for attaining promotion and success. This term is familiar to any Chinese whose DNA is built to aim high in order to achieve goals. Because *"gao"* also sounds like the word for cake, eating it on Double Ninth Day is a way to ensure that success lasts doubly long.

the heights of huan jing

When Huan Jing was a young boy, he was haunted by a recurring nightmare. The boy dreamed that the God of Plague would one day rise to destroy his village. Understanding his dream to be a prediction of imminent disaster, his life's mission became clear as he consulted the village elder. The wise elder told Huan Jing that no one had ever before escaped the God of Plague's wrath. However, Huan Jing was undeterred and resolved to take action.

Soon Huan Jing left his home to seek the advice of a wise monk. He scaled mountain ranges, crossed rivers, and wandered through forests until a bird led him to the gates of an old temple belonging to the priest Fei Zhangfang. But Huan Jing couldn't gain entrance. All remained silent, with the doors solidly bolted, the air indifferent.

Several moons rose and fell while Huan Jing continued to wait outside the temple. With the test of time, the monk became convinced of Huan Jing's sincerity and determination and accepted the young man into his monastery. Fei Zhangfang taught Huan Jing the secret of mighty swordsmanship used to battle evil, and the student learned his craft well. Pleased with the progress, Fei Zhangfang rewarded his disciple with the demon-slaying Green Dragon Sword and advised the God of Plague would rise on the ninth day of the ninth moon. Faced with his destiny, Huan Jing understood his charge with even more clarity.

Huan Jing immediately prepared to return home. As he set off, Fei Zhangfang handed Huan Jing a packet of protective charms containing dogwood, a green bag containing zhuyu grass, and a gourd of chrysanthemum wine. The priest advised Huan Jing to give the charms to the villagers and send them to the high ground.

The instant Huan Jing arrived home, he warned the villagers and arranged for their departure to safety. Then he waited patiently for his nemesis to arrive.

When the God of Plague suddenly arrived and spied the villagers, he roared up the hill to prey on them. However, just as suddenly, his senses were assaulted by the smell of dogwood, zhuyu grass, and chrysanthemum wine. Repulsed, he quickly turned toward the village.

In the moment that Plague descended, Huan Jing attacked with great vigor. Surprised by the fierceness of a mere mortal, the demon struck back relentlessly. Huan Jing likewise persevered, using Fei Zhangfang's teachings as encouragement. Huan Jing ultimately slayed the evildoer with the Green Dragon Sword. The God of Plague would never again return to spread suffering and death.

In honor of Huan Jing's success and tenacity, the Chinese acknowledge his courage and strength by hiking into the hills every year on Double Ninth Day.

nine-emperor god

Traditionally during Double Ninth, temple worship for the Nine-Emperor God occurs during the first nine days of the ninth lunar month. Incense sticks are lit. Flowers are laid, and joss papers are burned. Charms and trinkets are bought for the home and worn for protection and good luck.

The Nine-Emperor God is a composite entity of nine mythological emperors who are the sons of the Queen of Heaven. The god controls the future, health, and longevity.

The god represents many facets. In depictions, he appears on a lotus pedestal as an idol with four pairs of hands, one pair of which is clasped in meditation, and each of the other pairs holding the sun, the moon, a seal with Chinese characters, a spiked club, and a bow and arrow.

doubleheader

Insight and prophecy are staples to the Chinese. Fate is a classic Chinese belief. Destiny can be revealed through examining wet tea leaves, reading faces, or studying the mystical meaning of numbers. Double Ninth Day warrants a ritual gathering to fulfill aspirations. The number nine is considered yang for its masculine, solar-powered energy. "Nine" in Chinese is also a pun for "a long time," meaning that the duration of any occurrence endures twofold.

Based on old Chinese beliefs of cosmology, Double Ninth Day represents success and its longevity.

counting by numbers

Have you gotten a sense that repetitive numerical patterns and numerology are culturally significant to the Chinese? The Chinese holidays of Dragon Boat Festival (Double Fifth) and Double Seventh are another indication. Whether it's the numerical sum of a birth month and day or a set of numbers played on a keno card, the Chinese will subject the figures to critical analysis. Many numbers exhibit lucky qualities, and to help you level the playing field with Auntie Lao, here are nine numbers and what they signify:

• One—has yang qualities and is recognized as the first number; it represents unity.

• Two—represents duality and symmetry.

- Three—has positive traits because it sounds like "ever growing" in Chinese.

- Four—sounds like the word for death. The Chinese steer away from four, but double-four equals eight—a very, very good number.

- Five—the number of completeness. Many things Chinese come in fives: the five loves, the five colors, five directions, five elements, five metals. Five fingers and five toes.

- Six—a yin number that sounds like "good luck" in Cantonese.

- Seven—represents the seven days of the week, the seven festivals, and a completed journey after passing through the seven gates of the other-world that exists between earth and heaven.

- Eight—forever popular to the Chinese because it sounds like the word for prosperity and growth.

- Nine—connotes long life because of its similarity to the word for en-during. The Temple of Heaven in Beijing incorporates many nines in its design: The upper terrace of the altar is ninety feet tall; the plat-form of the temple has nine concentric circles of marble slabs. Even the marble balusters are in multiples of nine.

- All even numbers are yin, representing feminine energy ruled by the moon.

- All odd numbers are yang, connoting masculine energy ruled by the sun.

- Doubling certain digits, such as 5-5, 7-7, and 9-9, is considered good luck. Pairing items doubles the fortune.

part two

chinese special occasions

wedding

ceremony

You are sunlight and I moon

Joined by the gods of fortune

Midnight and high noon

Sharing the sky

We have been blessed, you and I

—*Opening Lyrics from*

"Sun and Moon," Miss Saigon,

Alain Boublil and Richard Maltby, Jr.

The Chinese believe the union of marriage joins the forces of dragon and phoenix, sun and moon, yang and yin, gold and jade. Marriage is the promise of continuing the ancestral lineage, joining two families and linking the present to the past and the future. Marriage is one of the most celebrated milestones in life. Taking a wife or husband is a rite of passage to full adulthood.

The old Chinese wedding processional complete with sedan chair and bright red lanterns no longer exists, but traces of an ancient tradition still linger. The continuity of generations past is evident in the way Chinese Americans have transformed the wedding into a multicultural affair.

The rituals of engagement and marriage are an ideal example. This is where a bride's "cookie" day endures alongside the bridal shower, a bachelor's weekend dovetails with the delivery of a bride's dowry, and the wedding ceremony of the West precedes the wedding banquet of the East.

It's no wonder that elaborate, monumental planning begins straightaway once a phoenix accepts a dragon's proposal of marriage. Life as she knows it will take new shape as she prepares to leave her family. And the old Chinese ways are at her disposal to invite life's blessings as she joins her groom to establish their very own branch on the family tree.

> *The Chinese mythological phoenix and dragon represent the essence of wife and husband. They're heaven's match of what yin is to yang, heaven is to earth, woman is to man.*

instinctive match

In prerevolutionary China, most marriages were arranged by the parents of a prospective bride or groom with the assistance of a well-meaning matchmaker. Frequently, the intricate negotiations for betrothal began at birth or when children reached school-age. By the time the children became adolescents, the matchmaker and the children's parents would have already promised the children to each other. It would be a matter of time before they'd marry in their teens.

An auspicious match would be based on the compatibility of a girl's and boy's astrological signs, birth dates, and family backgrounds. Since Confucian filial piety was widely practiced, the promised couple would obediently accept their parents' choice for wedlock sight unseen.

Today, whether a match is made by a well-connected auntie or an on-line matchmaking service, the annual Chinese almanac is still consulted to determine the most auspicious wedding date for a favorable beginning based on the couple's astrological signs and birth dates.

Auntie Lao says only the bold marry in the seventh lunar month, the ghosts' month.

According to Chinese astrology, the twelve animals either possess or lack affinity for one another based on their personality characteristics (see "Astrological Animal Years," pages 34–39). Naturally, some of the worst combinations are those at polar opposites on the astrological chart. For example, a persistent rat would annoy an independent horse, a hardworking ox could bore a whimsical ewe, while an authoritative tiger would frustrate a competitive monkey. Other mismatches are a discreet rabbit with a candid rooster, a demanding dragon with an introverted dog, and a decisive snake with a hesitant boar. Nevertheless, there are compatible Chinese astrological signs:

- The ambitious and energetic rat, dragon, and monkey
- The steadfast and determined ox, snake, and rooster

123 / weddings

- The humanitarian and idealistic tiger, horse, and dog
- The sensitive and artistic rabbit, sheep, and pig

Many Chinese consult a fortune-teller to evaluate a match by using the guidelines of Chinese astrology. Beyond the astrological animals, there are astrological overlays to consider because of the five elements (metal, wood, water, fire, and earth) associated with one's birth year. Such consideration could reveal that a fire-year monkey can be quite the opposite of a water-year monkey. Because a Chinese day is said to be only twelve hours long, the hour of birth is also considered, as the hours in a day correspond to the twelve astrological animals. The birth month and day are also determining factors for a couple's compatibility. But before analysis can begin, these dates must be converted into lunar calendar dates. Considering such absolute precision and skill, it's no wonder the highest respect is bestowed on the masters of Chinese divination.

eight characters

When all the lunar signs appear to be in alignment, the matchmaker gives the coupling a final test. The future bride's eight dispositions garnered from the year, month, day, and time of her birth are sent to the prospective groom's family. These Chinese characters are considered as unique to a person as a signature. To test the match, the girl's eight dispositions are written out and set upon the boy's ancestral altar for three days. If the days pass without misfortune, the match is considered positive. Next, the boy's eight dispositions sit on the girl's family altar for three days. When enough time has passed without any signs of protest from the otherworld, the respective characters of the couple's birth dispositions are written on red paper to serve as an announcement of impending marriage.

gift exchange

In early times in China when a match was made, a groom's family would send twelve betrothal gifts to the bride's family. The gifts included a pair of livestock, poultry, clothing, handkerchiefs, a pair of shoes (to symbolize harmony for the bride's younger unmarried brother), Chinese winter melon candies, dried longan fruit, pomegranates, long noodles, gold jewelry, red envelopes, and tea (the gift of engagement). In return, the bride's family would reciprocate with gifts that included lanterns, incense, candles, firecrackers, wine, ham, bird's nest, two jars with two goldfish each, gold chopsticks, pomegranates, and a portion of the groom's gifts returned. Then as now, when returning a portion of a gift package, the Chinese are implying that the giver's generosity was too great.

bride's cookie day

One month prior to the wedding date, the groom's representatives will deliver numerous foods and gifts to the bride's family. These items officially announce the engagement and help soften the loss of a daughter. According to old Chinese custom, a bride permanently leaves her own family after she marries. This gift-giving day is regarded as the bride's cookie day, when her extended family is invited to feast, celebrate, and bless the marriage. But, as with many of life's passages, her departure is tinged with the bitter and the sweet, for her engagement marks the first step away from her family home.

In Cantonese, the cookie day is commonly referred to as *lai beng*, which means gifting pastries. In fact, little ladies like Auntie Lao aren't shy about announcing their cravings for *beng* to singletons—giving them overt hints for a change of marital status. Today, the cookie day is typically an informal luncheon at the bride's family home.

For a large and elaborate cookie day event, the day begins in the morning with groomsmen delivering a whole roast pig (a symbol of purity) that they carry headfirst into the premises followed by an extraordinary assortment of fancy *beng*, or pastries, that contain rich fillings inside delicate pastry crusts. Popular fillings are sweet melon, fruit and nuts, coconut, lotus paste, and black and brown bean pastes. Some pastries are similar to moon cakes with an egg yolk center, while others are individual light sponge cakes in the shapes of flowers or are large almond cookies with pink watermelon seeds baked into the top. There are usually nine bridal *beng* varieties for the family to feast on—and for brown-bagging and eating at home for the next several days. Prospective grooms be aware: Given the size of many Chinese families, the number of cookies ordered can total into the thousands.

In displaying the *beng*, the pastries are placed in two large round red-lacquered Double Happiness cookie boxes along with dried longans, sugared coconut slices, and a few fresh juniper sprigs scattered on top. Oranges, dried lychees, peanuts, and dried red dates (known as jujubes) could also be scattered in the red cookie boxes to represent the five treasures of prosperity, wealth, longevity, many children, and peace.

Depending on the family's regional customs, the groomsmen may also deliver whole cooked chickens, Chinese black mushrooms, dried oysters, sea cucumber, and whiskey. Additionally, the bride's family will receive a ninety-nine-dollar dowry that the family can choose to return with a series of gifts for exchange with the groom's family. The future bride's mother and father could receive money in red envelopes for "shoes" and "pants," respectively, because in Cantonese the word "shoes" is similar-sounding to "harmony," and "pants" sounds like "fortune."

The groom could also give unmarried older brothers or sisters of the bride "pants" or "skirt" money in red envelopes for marrying over their birth order.

While the groomsmen wait, the bride's family gathers exchange gifts in appreciation of the groom's generosity. This gift collection includes boxes of *dim sum* dumplings and buns and an even number of each kind of bridal cookie. Moreover, the roasted pig's head, tail, and feet will be returned in the tray in which it was originally delivered with a *lai see* (red envelope) of nine dollars—a lucky number. Had the family received two chickens, one would be returned. Had there been four items, two would be returned. Sweet *tong yuen* (sweet rice flour balls) and a package of rock sugar are also included in the gift package. Not to be forgotten, the groom receives gifts too: a pair of pants, a pair of shoes, a belt, a new wallet containing ninety-nine dollars, and the family's wedding gift of Chinese jewelry. The future bride will often send her fiancé a watch as part of the jewelry collection. Eventually, the groom's gifts are boxed and scattered on the top with dried lychees, dried longans, peanuts, and dried red dates. Should the list of exchange items become overwhelming, it's also acceptable to substitute them with *lai see* money, with the items it's intended to represent written on the outside of the red envelope.

Today, some busy brides choose to forgo the cookie day in the family home. Instead, they deliver the *beng* or send a gift card that is redeemable for the pastries at a local Chinese bakery. Relatives who usually receive *beng* are the bride's grandparents, aunts, uncles, and cousins, and their respective families. A total of nine pastries (one of each variety) is often packaged and distributed to each family. Because the cookie day festivities are meant for the bride and her family, the groom and his family usually don't attend.

食 auntie lynn's engagement sponge cakes

makes 8 large cupcakes

These light yellow cakes are one type of beng served during a bride's engage-
ment. Easily made at home, the springy cakes are usually baked in 1¾-cup in-
dividual molds resembling flower blossoms that are available in Chinatown
kitchen and hardware stores. A regular muffin pan also suffices when the bak-
ing time is reduced to 12 minutes. An angel food cake ring or a 9-inch round
cake pan also works well. Thanks to this recipe, courtesy of my Auntie Lynn
Lowe, no one has to wait for a bride's cookie day to enjoy these delicious cakes.

8 large eggs
2 cups cake flour
1½ cups sugar
1 teaspoon baking powder
¼ teaspoon salt
½ cup vegetable oil
½ cup water
½ teaspoon cream of tartar
½ teaspoon lemon extract

1. Preheat the oven to 350°F. Grease eight baking molds and set aside.
In two bowls, separate the egg yolks from the whites.

2. Mix the flour, sugar, baking powder, salt, egg yolks, oil, and water together in a mixing bowl until blended and set aside.

3. In a separate mixing bowl, add the egg whites, cream of tartar, and
lemon extract. Beat the egg whites until stiff. Gently fold the egg whites
into the flour mixture until well combined. Pour the cake batter into the
greased molds.

4. Bake at 350°F on the lowest oven rack for 35 minutes, or until the
cakes are light brown and the tops spring back when gently pressed.

Place the cakes on a cooling rack and let cool completely in their molds. To easily release the cakes, insert a small spatula on the insides of the molds.

a bride's dowry

One to three days prior to the wedding, the bride's dowry is delivered to the groom's home. It often includes a Chinese hope chest containing domestic linens and bedding such as a silk comforter and matching pillowcases bundled together with a strip of red good luck paper. A set of dinnerware, complete with serving platters, soup tureens, a tea set, and chopsticks, is carefully packed for safe arrival. In addition to household furnishings, a monetary gift is enclosed to purchase other things for the new home. The final item in the dowry is a Chinese lacquered tray, known as the Good Fortune Tray, filled with lychees, longans, peanuts, dried red dates, and sweets such as sugared coconut, melon, lotus seeds, and ginger. This tray is similar to the octagonal New Year's Tray of Togetherness, which will shortly be used for the wedding tea ceremony.

Traditionally, a bride's dowry included the bride's trousseau, with red envelopes placed inside her shoes. But today the bride usually does not travel to her groom's home before the wedding day; thus her trousseau remains with her.

wedding attire

A Chinese bride's trousseau often includes the very best of East and West. Whether it's the traditional red silk *hong qua* wedding suit, the *cheongsam*, or the all-white wedding gown, a bride's wardrobe for the day requires a small army of bridesmaids and aunties to transport and manage it. A Chinese wedding includes multiple wardrobe changes throughout the festivities, not only to heighten the celebration, but also to allude to the bride's family's wealth.

From the wedding ceremony through the reception, Chinese brides don the Western-style white gown with a train and veil. But once the banquet begins, she will change from white to Chinese red, the color of happiness. Since white is considered a color of mourning, Chinese brides will avoid wearing only white all wedding-day long.

The traditional *hong qua*, meaning "red suit," has been considered the Chinese wedding gown since its creation for the imperial family in the Yuan dynasty (1279–1368 C.E.). This matrimonial suit consists of an intricately embroidered jacket of phoenixes and dragons—the symbols of the bride and groom—in silver and gold threads. A long pleated red silk skirt embroidered with flowers completes the ensemble. The Chinese wedding *qua* also includes an elaborate headpiece with red pom-poms and tassels—a Chinese counterpart to the veil—but it's not often worn today.

Few brides and grooms can resist the shapely and form-fitting *cheongsam*, which simply means "long dress." It's popular as feminine attire for a wedding banquet or other more formal social occasions because it transmits elegance. The salient characteristics that distinguish a *cheongsam* are a crossover-opening, high mandarin collar, a sleek fit, and slits up to the thigh. It is often sleeveless or has short capped sleeves. Formal versions are made of silk brocade or jacquard or are embellished with embroidery, intricate beadwork, and sequins. It envelops the body. Decorative knotted closures called frogs tease the eyes. Formal wear dictates a high stiff neck. Low necks are for every day. *Cheongsams*, also known as *qipao*, may be bought off the rack or custom-made by Chinatown tailors and seamstresses. Red, pink, gold, and silver are especially popular palettes for brides, as those colors represent happiness and prosperity. Frequently, the bride will change into several *cheongsams* throughout the course of the wedding banquet. Needless to say, each transition dazzles the eyes.

The wedding attire of a Chinese American groom is usually a tuxedo for the entire day. A traditional Chinese gentleman's robe is occasionally worn, but this practice is becoming less common in the West. Traditional robes are created with a subtle pattern on dark silk. A long red sash is draped over the shoulder and tied at the waist. This sash converts to a *mao dai,* a sling that a Chinese mother uses to carry a baby on her back. A round black hat with a red tassel completes the groom's traditional wedding attire. For a contemporary fashion statement, grooms now opt to wear a tasteful mandarin-style suit topped with a medium-collared jacket. The groom wears the ensemble during the wedding banquet so he can share the bride's flair for fashion.

baubles, bangles, and bands

If you observe closely, you might spot security guards at a Chinese wedding banquet. This is not because the bride and groom are considered new royalty, but because the bride is wearing the family's jewels. Chinese brides often receive gifts of twenty-four-karat gold jewelry that are presented by her close relatives to sweeten her departure from the family, and by the groom's relatives to extend a hearty welcome. At the banquet, a bride wears all the jewelry she has received, for yet another statement of prestige. Be assured that all the little aunties will notice each and every piece.

A bride's jewelry chest could contain a collection of twenty-four-karat Chinese gold necklaces, bangles, bracelets, and matching rings.

Pearls are favored because they're said to be the essence of the moon, signifying feminine yin beauty and purity.

Jade, known as *yuk* in Cantonese, is considered a celestial jewel that bridges heaven and earth. It's a symbol of protection and great virtue. Jade is believed to possess magical qualities and supernatural powers. It's said love can be invited by wearing a jade piece in the shape of a butterfly, which is a symbol of love. When love arrives, a jade piece is given to formalize the union. Hence traditional oval or rectangular jade rings are often worn by Chinese husbands. Nowadays, Westernized gold and platinum bands are more popular, and brides may want diamonds in the wedding rings but may opt to wear a simple band every day.

Chinese gold jewelry is generally sold by weight according to the daily market price. Prices for gold jewelry usually start at a few hundred dollars depending on weight, but a unique design can be considered priceless. Always be prepared to negotiate. Auntie Lao will wrestle down the initially quoted price by a third until she's satisfied. Numerous pieces warrant a quantity discount. The refined talent of a haggler and a strong stomach always serve the jewelry shopper well. Auntie Lao's greatest pride is in gaining her jeweler's respect with a price well earned. It's worth tomorrow's price of gold.

Love Lost, Love Found

Long ago in China, there lived a young girl of thirteen years who longed to study the Chinese classics. But her desire was uncustomary because it was a time when girls were confined to their homes and not allowed to venture alone in public, much less study alongside males. For her own protection, her parents refused. Their decision broke the young girl's spirit, and soon she began to waste away. Alarmed, her parents sought the help of every healer nearby, but all treatments proved futile.

One day, a strange man passed by the girl's home and learned of her parents' plight. He claimed he could mend their daughter, and so they anxiously invited him inside. As he crossed the threshold of the house, he tossed off his robes to disclose his true identity: He was actually their daughter disguised as a young scholar. Elated to see their daughter revitalized, her parents allowed her to leave home to study as long as she promised to return home at age sixteen to be married. She agreed and departed in disguise as a boy, to pursue her hungered-for education.

At school, the masquerading young girl befriended a fellow classmate who was assigned as her roommate. The senior classmate noticed the young student's naïveté, took pity on her, and took her under his tutelage. During the course of the three years of schooling, the two became inseparable. When it was time to return home, she could no longer continue the charade and had to reveal the truth. Writing as her true self, she divulged her secret identity and declared her love for him. The classmate, who already secretly held an inexplicable attraction for the young student, was elated by the news and immediately set out to pursue his newfound love.

Upon arriving at the girl's village, the classmate learned that his beloved had already been betrothed to another. Realizing that life without her would be too torturous, he succumbed to a broken heart within days. On her wedding day, when she learned of her true love's passing, the girl insisted that her wedding palanquin detour to his grave site. Still dressed in red bridal dress, she bent down to embrace the mound. At that instant, the earth beneath her cracked open, and she willingly entered its embrace. Not long after, the two lovers returned to earth as a pair of fluttering butterflies, symbolizing earth's purest love—unchanged and unhidden.

the chinese rituals of matrimony

In old China, the wedding day was the first time a bride and groom would set their eyes on each other. But a mere peek at the bride couldn't be had until the groom lifted the red silk off her veiled face. Many of the former rituals are now obsolete, yet many modern-day customs have evolved from the following five nuptial practices, all reflecting a romantic past.

ritual one: the bride's departure

Traditionally, on the couple's wedding day, a bride was fully attired from head to toe in classical Chinese red as she prepared to leave home. In bidding a final farewell to her family, she served tea to her elders and honored her ancestors at the home altar by bowing three times and burning incense. Afterward, the groom's entourage arrived to take her away in a curtained sedan chair made of bamboo. In bold Chinese characters, banners announced the couple's family names, and the processional was accompanied by red and gold lanterns and clanging cymbals and gongs.

Today, on the morning of the wedding, a groom and his ushers arrive at the home of the bride in a limousine or motorcade rather than a sedan chair. But before gaining admittance into her home, the groom is badgered into paying lucky money to the bride's protectors, her bridesmaids. Given in a red envelope, the total amount should contain all nines ("nine" is a Chinese synonym for "everlasting"). Whether it is $99.99 or $999.99, the amount is a sign of a groom's worthiness. This well-deserved lucky money is shared among the bridesmaids.

Finally, when the groom and his party gain entrance into the bride's home, a light meal—considered the bride's last one at her home—of *dim sum*, chicken, roast pig, prawns, fish, and *full jook tong* (dried bean curd soup)—is served. Now dressed in red Chinese wedding attire, the bride and her groom, *jum cha*, serve ceremonial tea to her parents and elders as a parting gesture.

In the old days, as the bride stepped beyond her family's front door, she did not stop to linger, or positive energy and her family's fortune would flow out through the door. Wearing her ornate headdress and veil, she was carried to the sedan chair by the matchmaker so that her feet would not touch any evil spirits on the ground. Then the bride threw a fan into the air as a wish for a successful marriage. Winter melon candies were tossed as a symbol of throwing out bad old habits and sweetening the sorrow of leaving home. During the sedan-chair journey, the bride held an apple, a symbol of peace, to bring tranquillity to her groom's home.

The bride cannot return to her family home until the third day after the wedding rites and should be accompanied by her husband. On the tenth day and thereafter, she may visit without him, for she is then regarded as belonging to another family.

All daughters must leave home,
But few have your luck.
Your mother-in-law's house is a rich fish
*   pond.*
You'll doff your cotton tunics,
And wear only silk.
Yes, you're entering the Dragon Gate of
*   Fortune.*
You'll ride to the sky on a white crane.
*   —Ruthanne Lum McCunn, The Moon Pearl*

ritual two: the bride's arrival

Traditionally, when the bride arrived at the groom's home, firecrackers were lit to ward off evil spirits and mark the happy occasion. Before the bride stepped down, the sedan chair was passed over a hot pot of charcoal to burn away any evil spirits, and the groom shot three arrows into the sky to scare away any nearby demons. As the palanquin was set down, the bride emerged and stepped over a saddle, a symbol of stability, and was presented with a tangerine, a symbol of good fortune.

Today, no coals or arrows are required because the wedding party will arrive directly at the wedding ceremony site, where the bride will change into her Western-style wedding gown and veil.

ritual three: the wedding altar

In old China, the Chinese matrimonial rites included a red sash to signify the union of the couple. Delicate nuptial wine cups were connected by red string. Once the couple exchanged cups and crossed arms to drink the wine, the alliance of matrimony was complete. The sharing of wine completed the circle for creating a harmonious marriage.

Today, wedding rites in a Westernized service are often performed by a cleric or officiant. Commonly, a bride in full wedding attire is escorted down the aisle to the altar in accompaniment of music. Readings may be given and unity candles may be lit. Amid bouquets of fresh flowers and busy photographers, vows are exchanged.

ritual four: paying respects at the family altar

Traditionally, when Chinese rites were conducted in the groom's home, the newlyweds honored his family and ancestors by bowing three times at their home altar. Then firecrackers were lit to signal the end of the ceremony.

ritual five: meeting the family

The final traditional wedding ritual of serving ceremonial tea to the groom's family still is widely practiced today. A wise auntie often oversees the tea ceremony to help organize the family and coordinate the tasks. The tea is served to family members in descending order, from the oldest to the youngest: grandparents, parents, aunts, uncles, and older married siblings. The couple kneels on pillows and offers tea with two hands to their grandparents and parents. Tea is then served to the other relatives with the couple standing. In return, each relative offers a wedding present of jewelry or *lai see* (red envelope of lucky money). Special foods are arranged in a Chinese lacquered tray that has been sent with the bride's dowry and placed on the tea ceremony table. These items include oranges for luck, sticky steamed cake called *gao* representing advancement in work, candied fruits such as kumquats for prosperity, and sugared lotus seeds for many sons.

the wedding banquet

The wedding banquet is considered the main event of the celebration. To the Chinese, sharing abundant, elaborate foods and drink in times of family unity is the epitome of harmony. Some couples opt to combine the wedding reception with the banquet depending on the flow of the wedding day activities. Many of the Western customs—tossing the bride's bouquet, throwing the garter, cutting the wedding cake, and the couple's first dance—are also conducted at the Chinese banquet.

When arriving for the banquet, guests first check in at the reception table and sign the red silk tablecloth that doubles as the guest registry. The silk cloth, which the couple can later frame and hang in their new home, will be imprinted with the Double Happiness characters or have a dragon and phoenix embroidered on it. Often there will be greeters from the family to facilitate seating arrangements for guests. At some banquets, seating is designated for the entire guest list, while at others, the hosts may choose to reserve tables only for close family and friends and leave the remaining tables for open seating.

At the reception table, family members are given corsages and boutonnieres to differentiate their familial status. Deciding who receives flowers is at the discretion of the family, but usually the members include grandparents, great-aunts and -uncles, the couple's parents, aunts, uncles, siblings, cousins, nieces, and nephews. The Chinese assign corsages and boutonnieres according to generation: Relatives who are older than the new couple must wear flowers, while those who are the couple's peers have an option. This whole floral decision is often dictated by the groom's parents, because typically they incur the expenses for the wedding banquet. The bride's parents usually assume expenses for the ceremony and reception.

Because the first hour of a banquet is considered a time for guests to socialize and settle in, the master of ceremonies presents the happy newlyweds to the guests during the next hour. They enter the room regally, the bride resplendent in her traditional red silk suit and all her Chinese jewelry wedding gifts. The crowd applauds and guests begin to lightly tap their chopsticks on their drink glasses. They will not allow the couple to sit until they witness a kiss. The chopsticks are insistent. The *tink-tink-tink* wears the couple down until they surrender to the command. Only brief respites are allowed before the crowd repeats their chopstick action again throughout the evening.

Between the couple's kisses, the master of ceremonies makes a series of introductions and speeches about the momentous occasion. The groom's family is introduced first, followed by the bride's. During the course of the accolades, the host family may choose to make a series of charitable contributions to various benevolent associations and social organizations with which they are affiliated. This act stems from the Chinese belief that giving good fortune is an invitation for its return.

Dinner is preceded first by the best man's wedding toast, during which the guests are asked to stand and also raise their drink glasses to the couple. Then there may be a heart-pounding lion dance, where flirtatious lions will prance, bow, and bat their eyes at the newlyweds. Once the playful lions exit the hall, the evening is left to the deft coordination of the banquet room captain, who synchronizes the waitstaff with the industrious kitchen while dinner is served.

banquet cuisine

The wedding banquet menu is carefully selected for a culinary balance. To accomplish this feat, various cooking methods are adopted to achieve an array of flavors to awaken the palate. Cold balances hot, salt neutralizes sweet, boiled offsets fried, steaming equalizes braising. The presentation and color of the food is also important so that the combination of courses harmonizes with the eye as well as the tummy. Generally, the Chinese prefer lighter- over darker-colored foods, thus traditional Chinese wedding cake tends to be light in color. Devil's food cake would be a rare sight.

The banquet menu typically includes nine items because of the number's everlasting connotation. A sample menu of wedding banquet courses is below:

- Cold appetizer platter with a variety of meats and seafood such as roast suckling pig, boiled ham, beef tongue, soy sauce chicken, and pickled jellyfish
- Shark's fin or bird's nest soup—considered a Chinese delicacy
- Whole roasted chicken—a symbol of the phoenix
- Whole squab—a symbol of peace
- Lobster or shrimp—signifying the dragon
- Stir-fried vegetables containing water chestnuts—connoting progeny
- Sweet-and-sour or other red-colored dishes for good luck
- Peking duck with steamed buns and hoisin sauce
- Whole steamed fish—symbolizing abundance
- Fried or steamed rice
- A pair of desserts—wedding cake and sweet lotus seed soup to wish the blessing of successive sons.

> Auntie Lao says life's happiness comes in pairs.

the wedding night

When the wedding rituals are complete, the last detail of union lies in the wedding chamber. Traditionally, a couple's bed is decorated in lucky red embroidered pillows and silk bedding. Grains, gold coins, dried fruits, and lotus seeds are scattered on the nuptial bed to encourage a fulfilling life of work, wealth, and family. Before the couple retires to their chamber, toddler boys are recruited to frolic atop the bed as an additional blessing for many sons. At the end of the evening, all that remains is up to the sweet lotus seed soup.

 sweet lotus seed soup

serves 8

This warm dessert is considered the bride's soup because lotus seeds are symbolic of continuous lineage. This recipe was garnered from the Empress of China Restaurant in San Francisco, a long-standing doyenne of the Chinese banquet, and has been adjusted for a family of eight rather than eight hundred dining guests.

> **½ cup dried lotus seeds**
> **½ cup dried lily bulbs**
> **8 cups water**
> **¼ cup sugar, plus more to taste**
> **¼ cup canned evaporated milk**
> **6 tablespoons coconut milk**

In separate bowls, rinse and soak the dried lotus seeds and dried lily bulbs in warm water. Soak overnight. In a pot, combine the 8 cups water and the lotus seeds. Bring to a boil and simmer for 45 minutes. Add the lily bulbs and simmer for 45 minutes. Add the sugar and stir. Add the evaporated milk and the coconut milk. Bring to a slow boil. Sweeten with additional sugar to taste. Remove from the heat and serve warm.

the wedding planner

Once *the* question is popped and it's met with a favorable response, the countdown to the wedding day of your dreams begins. The wedding planning calender below integrates traditional Chinese practices with current American customs.

Activity	Suggested Timing
Consult the Chinese almanac for an auspicious wedding date. Visit a Chinese fortune-teller.	10 to 12 months in advance
Select your engagement ring, if it hasn't already been presented.	10 to 12 months in advance
Meet with both families for mutual preferences on the wedding style, dates, location, number of guests, and payment of expenses.	10 to 12 months in advance
Take an engagement photograph for a newspaper announcement.	10 to 12 months in advance
Select the wedding party—bridesmaids, groomsmen, ushers, ring bearer, flower girl, and other people who will participate in the ceremony.	10 to 12 months in advance
Celebrate your engagement with family and friends.	10 to 12 months in advance
Reserve the ceremony, reception, and banquet sites. Venues in major metropolitan	10 to 12 months in advance

cities may require booking 12 months in advance.	
Begin locating a caterer, baker, florist, photographer, videographer, and musicians (band, performers, or DJ).	10 months in advance
Determine who will preside over your wedding service. Research requirements for premarital counseling, as requested by some religious denominations.	10 months in advance
Confirm your primary suppliers— caterer, baker, florist, photographer, videographer, and musicians (band, performers, or DJ).	8 months in advance
Select and order your wedding dress, veil, and headpiece. Begin search for shoes.	8 months in advance
Select traditional Chinese wedding dresses (hongqua and cheongsams) that can be custom-made, purchased off the rack, or rented from Chinatown's special-occasion dress stores.	8 months in advance
Shop for bridesmaid and flower girl dresses.	8 months in advance
Research honeymoon destination.	6 to 8 months in advance
Gather invitation list and addresses.	6 months in advance
Discuss details with all wedding suppliers—caterer, florist, baker, photographer, videographer, and musicians (band, performers, or DJ).	6 months in advance

Activity	Suggested Timing
Confirm final requests and requirements due from all suppliers.	
Book limousine service.	6 months in advance
Order wedding invitations, enclosure cards, and maps.	6 months in advance
Select and order wedding favor items.	6 months in advance
Reserve hotel rooms for out-of-town guests.	6 months in advance
Recruit a responsible wedding coordinator.	6 months in advance
Make honeymoon travel arrangements. Apply for passports if necessary.	4 to 6 months in advance
Order wedding rings.	4 to 6 months in advance
Secure rentals—tent, tables, glassware, if needed.	4 to 6 months in advance
Arrange for the rehearsal dinner. Assemble guest list if including more than wedding party.	4 to 6 months in advance
Reserve a room for the wedding night.	4 to 6 months in advance
Schedule first wedding dress fitting.	4 to 6 months in advance
Assign dresses to bridesmaids and flower girls.	4 to 6 months in advance
Select bridal accessories.	4 to 6 months in advance
Select and reserve formal wear for groom, groomsmen, ushers, ring bearer, and father of the bride.	4 to 6 months in advance

Activity	Suggested Timing *(continued)*
Review vows, readings, and music with presiding cleric or officiant.	3 months in advance
Check status of bridesmaid and flower girl dresses.	3 months in advance
Inquire about marriage license requirements. Make a doctor's appointment if a blood test is needed.	3 months in advance
Sign up for bridal registries.	3 months in advance
Give required number of Chinese beng (engagement cakes) to groom's family.	3 months in advance
Finalize the invitation list. Address and prepare invitations. Mail 4 to 6 weeks in advance.	2 months in advance
If bride is taking the groom's name, obtain name change forms from official agencies: Department of Motor Vehicles, Social Security Administration, Voter Registration, etc.	2 months in advance
Make hair- and makeup–stylist appointments.	2 months in advance
Plan bachelor and bachelorette parties to take place a few weeks before the wedding.	2 months in advance
Attend bridal shower. Present gifts for shower hostesses.	6 weeks in advance
Schedule second dress fitting with wedding shoes.	6 weeks in advance
Begin assembling wedding favors.	6 weeks in advance
Prepare and print wedding ceremony programs.	6 weeks in advance

Activity	Suggested Timing
Invite bride's family and friends for Bride's Cookie Day.	6 weeks in advance
Bride's Cookie Day—Groomsmen deliver to the bride's home engagement cakes or beng, a whole roast pig, and exchange gifts. The bride's family sends exchange gifts to the groom's family.	1 month in advance
Reconfirm arrangements with officiant, caterer, baker, florist, rehearsal dinner, reception, and banquet venues, limo service, photographer, videographer, musicians (band, performers, or DJ).	1 month in advance
Apply for marriage license (timing depends on licensing state).	1 month in advance
Purchase gifts for each other, the wedding party, and bride's and groom's parents.	1 month in advance
Schedule final fittings for bride's outfits and groom's attire.	1 month in advance
Get immunizations for honeymoon if necessary.	1 month in advance
Pick up wedding rings.	1 month in advance
Estimate guest attendance based on RSVPs. Contact invitees who haven't responded.	3 weeks in advance
Provide final guest count to caterer and banquet facilities. Create seating chart if using designated seating.	2 weeks in advance
Prepare menu cards or table cards if using.	2 weeks in advance

Activity	Suggested Timing *(continued)*
Reconfirm rehearsal dinner arrangements and final guest count.	2 weeks in advance
Reconfirm all honeymoon arrangements and reservations.	2 weeks in advance
Provide a "shot list" for the photographer and a music list for the musicians.	2 weeks in advance
Prepare a list of relatives to serve during the wedding tea ceremony—grandparents, great-aunts and -uncles, parents, aunts, uncles, and married siblings. Select a tea ceremony coordinator who can assist with the ritual. Gather tea ceremony items: tea set, tea, fortune tray, pillows for kneeling.	2 weeks in advance
Pick up printed thank-you cards.	2 weeks in advance
Attend bachelor and bachelorette parties.	2 weeks in advance
Deliver bride's dowry to groom's home.	1 week in advance
Create wedding day schedule.	1 week in advance
Create list of responsibilities and give to attendants and groomsmen.	1 week in advance
Pack for the honeymoon. Delegate luggage responsibilities to a groomsman.	1 week in advance
Notify all wedding participants of the rehearsal schedule and rehearsal dinner date, time, and location.	1 week in advance

Activity	Suggested Timing
Pick up wedding dress and store it in a safe place.	1 week in advance
Pick up groom's and attendants' attires.	1 week in advance
Coordinate bride's wedding day trousseau, including Chinese attire and accessories.	1 week in advance
Give a bridesmaids' luncheon, distribute bridesmaids' gifts, and assign wedding day responsibilities.	1 week in advance
Arrange for the transportation of gifts.	1 week in advance
Distribute wedding day schedule to all involved. Remind everyone of photo times, assembly locations, etc.	1 day prior
Take time for relaxation—a massage, a manicure, or a warm bubble bath.	1 day prior
Attend rehearsal and rehearsal dinner. Be prepared to toast your party.	1 day prior
Retire to bed early.	1 day prior
Attend hair- and makeup-stylist appointments.	The wedding day
Distribute "new address" cards and announcement of bride's name preference.	The wedding day
Let yourself be swept off your feet when the groom arrives with his lucky money. Eat a light meal and keep hydrated. Serve tea to bride's family	The wedding day

Activity	Suggested Timing *(continued)*
before departing for ceremony site.	
Give the best man the bride's ring, marriage license, and necessary payments. Give the groom's ring to the maid of honor.	The wedding day
Let your wedding coordinator and wedding party take control.	The wedding day
Dazzle and delight—it's your day!	The wedding day

complete

month

Auntie Lao's

Advice for Expectant Mothers

Don't eat bananas or your baby will have big ears.

Don't eat pineapple or your baby won't have smooth

skin. Don't eat ginger or your baby might have more than

five fingers. Don't visit the zoo or your baby will look like a

monkey. Don't sew while sitting on the bed or your baby

could have a harelip. Don't reposition your bed or it will

change your feng shui. Don't move any furniture in

your house. No heavy lifting or stretching.

No hammering nails.

Don't eat lamb.

Don't eat shell–

fish. Don't eat

watermelon.

Don't go to

the cemetery.

There are few experiences in life that deliver the joy and hope a new mother and child bring to a Chinese family. It may not be politically correct, but the Chinese can't help but beam with delight when it's a boy because a son perpetuates the family lineage through his name. This doesn't negate the fact that a daughter has her special place in the world. Nevertheless, whether boy or girl, a new Chinese baby is reason enough for celebration with a myriad of symbolic foods and rituals that invites a child's good fortune.

the sitting month

The first month of a baby's arrival should be spent "sitting" at home so that the new mother and infant can gain their full energy and strength. Called *chau yuet* in Cantonese and *zuo yue zi* in Mandarin, it's a time when the new mother shouldn't leave the house, shower, wash her hair, eat fresh fruit, or drink cold beverages.

The Chinese believe that a new mother is *leurng*, or cool. Since her life energy, or *qi*, is considered full of yin following childbirth, efforts are made to rebuild her body by restoring the internal balance of yin and yang. Symptoms of being *leurng* are anemia and light-headedness. When new mothers are *leurng*, they are highly susceptible to getting *fung sup*, which is like a cold wind entering the deepest crevices of an old woman's bones. Once the wind gets into the bones, it's considered irreversible, like arthritis. It makes little ladies like Auntie Lao shiver with dread. One critical agent for invigorating a new mother's strength is ginger, the heat-generating aromatic that warms the body.

The opposite of leurng is yeet-hay, which means "hot breath" in Cantonese. This term refers to a bodily imbalance of too much internal heat, or yang. Typical symptoms of yeet-hay are canker sores in the mouth, sore throat, and abnormally bad breath. Simple yeet-hay remedies used in my home were avoiding spicy or fried foods, drinking boiled carrot water with sugarcane, and slurping sweet scrambled egg soup.

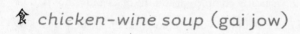 chicken-wine soup (gai jow)

serves 6

To restore a woman's health after giving birth, new moms are given this traditional Chinese dish upon returning home from the hospital. Chicken-wine soup, typically eaten during the sitting month, is believed to help wash away the remnants of childbirth and increase body heat. All who visit the new baby are also invited to celebrate the birth by eating this potent wine-flavored chicken soup that is lean and strong. Chicken-wine soup is typically included on the mun yurt (newborn's "full month" milestone) and Red Egg and Ginger Party menus.

10 to 12 Chinese dried black mushrooms
20 to 30 dried tiger lily buds
2 cups dried wood ears (*mook yee*)
One 2- to 3-pound chicken
Nonstick cooking spray
4 to 5 large pieces of ginger, the size of a walnut, peeled and sliced
6 to 8 dried red dates (*jujubes*)
1 to 1½ cups Sancheng Chiew rice wine or gin
2 to 3 rock sugar pieces
½ cup skinless raw peanuts (optional)
Salt

1. Place the dried mushrooms, tiger lily buds, and wood ears in separate bowls. Fill each bowl with enough hot water to cover and soak for approximately 2 hours, or until soft. Wash and rinse the individually soaked ingredients until clean. Trim off any hard stems or pieces. Set aside.

2. Cut the chicken into 2- to 3-inch pieces by cutting the drumsticks into two pieces, separating the wings into two pieces, and cutting the breast into three sections.

3. Lightly coat a large pot with nonstick cooking spray. Brown the chicken over medium heat. Add the ginger slices and continue browning. When the browning is nearly completed, add the black mushrooms, tiger lily buds, and red dates. Add hot water to cover all of the ingredients by at least 2 to 3 inches. Bring to a boil and simmer for 25 minutes. Add 1 cup rice wine and simmer for another 25 minutes. Add the wood ears, rock sugar, and peanuts (if desired) and simmer for 10 minutes.

4. Prior to serving, add salt and additional rice wine to taste. Reheat and serve.

食 auntie ruby's black vinegar pigs' feet

serves 10

During the first sitting month, black vinegar pigs' feet are served around the clock along with chicken-wine soup. This dish is deep brown in color with a rich, sweet sauce of ginger and brown sugar. The meat is simmered to such softness that it dissolves in the mouth. Black vinegar pigs' feet are believed to hold restorative powers and to build up a mother's breast milk. This dish is also served during the baby's mun yurt ("full month" milestone) and Red Egg and Ginger Party. Below is my Auntie Ruby Young's version, and it's considered my family's finest eating whether sitting or standing.

6 to 8 pigs' feet, cut into quarters
Five bottles (20.3 fl. oz.) Koon Chun Sweetened Black Vinegar
One bottle (20.3 fl. oz.) Koon Chun Black Vinegar
6 cups water
6 to 7 large pieces of ginger, peeled

6 rectangular Chinese brown sugar bars
6 hard-boiled eggs, peeled (optional)

1. Rinse the pigs' feet and put them in a large stockpot with water. Parboil, drain, rinse, and change the water. Repeat the parboiling process two more times. After the third boiling, rinse the pigs' feet under cold water.

2. Combine the sweetened black vinegar, black vinegar, and water in a large nonreactive pot (e.g., enamel). Add the ginger, brown sugar bars, and pigs' feet to the pot. Bring to a boil. Simmer for 1 to 1½ hours, until the pigs' feet are soft. Stir occasionally. Skim the fat off the surface if needed. Add the hard-boiled eggs (if desired) to the sauce right before serving hot.

Auntie Lao's advice for new mothers and infants is to stay cloistered at home for thirty days before venturing out in public. Consider the time as an incubation period for mother and child to gain full strength and health. Here's a list of Auntie Lao's do's and don'ts after returning home from the hospital:

- Drink only warm liquids—no cold drinks.
- Eat lots of ginger.
- Don't wash your hair for thirty days.
- Don't shower.
- Take warm ginger baths.
- Don't go out in the cold.

full month

The first milestone in a child's life is reaching the ripe old age of one month. In fact, a Chinese baby's birth isn't officially announced until a full month of life has been completed. Known as *mun yurt* in Cantonese, this waiting period stems from the high infant mortality experienced in old China. With so much heartbreak, the Chinese proceeded cautiously and deferred a new baby's announcement to a "safer" time. This act of deference avoided tempting the gods to take away what had arrived auspiciously.

At one month of age, the child is welcomed with a celebration of firsts. There's the first bath, the first haircut, the first new outfit, and, finally, a new Chinese name. Some traditional families will shave the baby's head except at the top of the crown to remove the hair they considered was grown in the womb. This is to stimulate new hair growth. The fine baby hair is then bundled up and tied with red string and then stored as a keepsake.

The child is presented to the ancestors with spirit offerings of food and drink. The Chinese grandmother typically presides over the ritual by burning incense, presenting the child, and bowing three times in symbolic gesture of the three realms of heaven, earth, and the other world. During the ceremony, she peels a dyed-red hard-boiled egg and rolls it on the infant's skull, forehead, face, and body to invoke a life not only filled with good fortune but blessed with high rank.

At the baby's one-month birthday party, Ah Po gave him the Four Valuable Things: ink, inkslab, paper, and brush. The other chil-

dren had only gotten money. She put the brush in his right fist. Villagers and relatives praised the way he waved it about. Eating bowls of chicken-feet-and-sweet-vinegar soup and pigs'-feet-and-sweet-vinegar soup, they said good words for the future. Ah Po shaved her baby's head except for the crown, though it was only baby fuzz he had growing on his head. The house was ashine with lights and lucky with oranges.

—Maxine Hong Kingston, China Men

red eggs and ginger

A baby's full month is celebrated with dyed-red hard-boiled eggs and pickled gingerroot. Eggs are an auspicious symbol of fertility, birth, and life, and the color red is a symbol of happiness and good luck. Pickled ginger represents a family's strong, deep roots that their grandchildren perpetuate. Phonetically, the Cantonese words for pickled ginger, *sern gueng*, sound like the words "grandsons" and "gingerroot." For good luck, the ginger is pickled in dyed-red or pink brine. Combined with the restorative healing effects the aromatic ginger provides to a new mother, red eggs and ginger have long been considered the perfect pairing to celebrate a new Chinese baby.

Other dishes served during the *mun yurt* family dinner, which is usually held at home, are Chicken-Wine Soup (*Gai Jow*) (page 155); Auntie Ruby's Black Vinegar Pigs' Feet (page 156); roast suckling pig; sautéed vegetables; steamed buns; sesame balls, or *geen doi;* sweet rice flour dumplings, or *tang yuan;* apples; oranges; bananas; brown sugar blocks; fermented sweet rice wine pudding, known as *teem jow;* and steamed rice.

> *Auntie Lao says that when a new mother blows ginger breath on an infant's head and gently rubs it, the baby's head will grow rounder.*

what's in a name?

As part of the one-month initiation, a Chinese name is bestowed on the family's new addition. Selecting a Chinese name is part art form, part science. Through the combination of two simple but significant words (or sometimes using only one word), a Chinese name not only serves as an identity but expresses a family's aspirations and intentions for the child. The naming process can be quite rigorous and often requires a systematic approach to ensure that the symbolic meaning and the sound harmonize and complement each other.

family surname

A full Chinese name typically consists of three characters. The family surname always appears first, followed by the two-, or one-, character given name. Many surnames represent the Chinese village, district, or territory from which the family originated. Others reflect ancient dynastic emperors, rulers, and leaders from over five thousand years ago. The majority of Chinese surnames have only one character, such as Chan, Lee, Wang, or Yu. However, occasionally some are a two-character combination, resulting in surnames such as Au Yeung, Ow Yang, or Soo Hoo. In either case, the first rule of thumb when selecting a child's given name is that it complement the family's name in meaning, sound, and even visual balance of the number of strokes when the full name is written in Chinese characters.

given name

Because a Chinese name reflects a family's best intentions, the name is considered influential to a child's role and destiny. A child's given name is constructed to provide a solid first impression. Some names address the child's projected role in life, such as Family Pillar and Respectable Scholar. Others are endearing and cast a distinct image, such as Little Green and Smiling Plum Blossom.

Given names can be used to connect siblings. Names may or may not be assigned by gender. For example, a group of brothers could have unifying names as simple as Big Dragon, Middle Dragon, and Little Dragon, and a group of sisters could have unifying names such as Sweet Lotus, Sweet Rose, and Sweet Iris.

Selecting that perfect Chinese name sometimes requires the diligence of a drill sergeant. My advice is that two heads are better than one, so it would be smart to consult a wise, experienced elder. Below are broad parameters often used to enhance the naming process:

- Evaluate the child's birthday based on the eight Chinese characters consisting of year, month, day, and time of birth in order to find a name that's astrologically harmonious.

- Determine which of the five elements (metal, wood, water, fire, and earth) are complementary to the child's birth date and the family's surname.
- Consider the birth year's astrological animal and the animal's characteristic traits.
- Check the name for its yin or yang qualities, then its appropriateness to gender.
- Evaluate the number of strokes in the written name's Chinese characters to ensure a visually balanced grouping.
- Test the sound and intonation of the given name with the surname.
- Evaluate the name's connotation and meaning for the child.

> *"Ginger! I should have listened to my great-aunt! I should have given you the name she had suggested, 'Plain Water.' It was to calm you and tame your character. . . . She hired a fortune teller who told us that there was too much fire in you when you were born. . . . But, I didn't care. I liked the passion that fire signified! Your father and I named you Wu-Jiang, 'Wild Ginger,' because we loved the fire in you!"*
>
> —Anchee Min, Wild Ginger

generation names

Reverence for one's history, family, and ancestry is ingrained in Chinese society. Many traditional families' patriarchs establish a series of names according to generation. Thus these special names help to place one in a particular generation. Such names could follow a verse or phrase that a great-great-grandfather aspired to carry forth through his lineage. For example, in using a sixteen-word verse, an ancestor names all Generation I children (siblings and even cousins) by incorporating word one of the verse into part of the name. Generation II names incorporate word two of the verse, and Generation III names incorporate word three, and so on for future generations. Again, names may vary by gender. In an alternate method, each new generation is assigned its own name combinations that may or may not relate to previous generations. Nevertheless, generation names are a unifying reinforcement for entire families that may now be found residing around the globe.

When a Chinese woman marries, she does not relinquish her family surname but has the option of adding her husband's surname as the first character to her existing surname.

red egg and ginger party

As an introduction into the family's social circle, Red Egg and Ginger parties are given as the baby's official coming-out celebration. They're typically held when the baby is approximately one hundred days old and can take the form of a traditional banquet dinner, with multiple courses similar to a wedding or birthday banquet, or of a buffet-style lunch-eon—an increasingly popular party type in America. Often the child's parents and paternal grandparents serve as hosts. As with other celebra-tions, once a party date and restaurant are chosen, invitations are sent, and decorations and favors are arranged if preferred.

> While growing up, I remember family visits in which new parents came calling with their new baby. They would bring red-egg-and-ginger "announcement" packages that included the obligatory dyed-red hard-boiled eggs and pickled ginger, chicken-wine soup, black vinegar pigs' feet, and steamed buns. This was how a new baby was introduced when an open-door policy was still a way of life.

goo-goo and ga-ga
baby gifts

Any gift selected for an American-style baby shower is appropriate and appreciated for a Red Egg and Ginger Party: baby clothes, receiving blankets, toiletries, and toys. All apply. But should you find yourself browsing in Chinatown, there are traditional baby gifts that aren't stocked in the typical shopping mall.

tigger or pooh?

According to Chinese folklore, the king of all beasts, the tiger, possesses protective powers to keep children safe. As a motif for infants, tigers are as popular as Mickey Mouse. For gifts, look for tiger hats, tiger clothing, and tiger slippers. Note that on the baby slippers, the tiger's eyes are always wide open and watchful because they divert danger from wee ones so that they can stay steady on their feet. Oftentimes, Chinese grandmothers will sew gold, silver, and jade charms onto infant tiger hats for their grandbabies' good luck and protection. Tiger-motif clothing can be found in Chinatown apparel and souvenir stores.

baby jewels

Nestled inside Chinese jewelry stores are several small treasures for new babies. Close relatives often give miniature twenty-four-karat-gold bracelets or bangles for baby girls to wear. A necklace with a feminine flower charm or jade heart is also a popular gift. For boys, necklaces with a gold coin, fish, or jade pendant are equally symbolic. Another popular gift many jewelry stores carry is gold charms representing astrological animals of the lunar calendar.

Auntie Lao says a gift of jade calms a child's heart.

reaching for red

Lucky money, *lai see* or *hong bao*, is the universal gift for all important occasions, and a Red Egg and Ginger Party is no exception. Red envelope contents are bountiful and intended to jump-start a good luck life, and the riches are often deposited into a college fund. At the celebration, some guests may hold a red envelope within the baby's reach to watch in amusement if the baby will clasp it. How hard the baby hangs on to the red packet could be a sign of a child's prosperity. Napping babies often have envelopes tucked into their receiving blankets. To ensure that your red envelope doesn't get lost, enclose it in a greeting card, or simply write your name on the back of it. In any case, the family will acknowledge your generosity.

> When a baby reaches one hundred days old, Auntie Lao initiates the little one with a new high chair and a chicken drumstick to chew on.

the fortune tray for work and play

On the baby's first birthday, the Chinese try to predict a child's future profession by playing a game of fortune-telling. They place the baby in a high chair, then place several items on the eating tray. Each item suggests a line of work. For example, a pencil and a book mean the child will become a scholar or educator; a calculator suggests a businessperson; scissors represent a dress designer or hairstylist; a paint brush means an artist; a hammer and a screwdriver imply a carpenter; cars and trucks suggest a mechanic; a stethoscope suggests a doctor. What object the child reaches for determines the profession. Unfortunately, the true result will require some time to reveal itself.

> On her grandchildren's first day of school, how does Auntie Lao plant the seed of success? By sneaking a green onion and a nickel into the pocket of the unsuspecting child. The green onion, or *choon* in Cantonese, is similar-sounding to the word for smart, and the nickel suggests success.

planning a red egg and ginger party

The good news is that planning a Red Egg and Ginger Party takes less planning and effort than a wedding. The bad news is that it does require some attention. Since Red Egg and Ginger parties tend to be informal affairs, many of the preparations can be accomplished during the baby's naptime or delegated to proud new uncles, aunties, and cousins.

Activity	Suggested Timing
Select party date and time and reserve the restaurant. Begin compiling the invitation list.	2 to 3 months in advance
Arrange for invitations.	2 months in advance
Select and order favors (optional).	2 months in advance
Mail invitations (RSVP options can be via phone or e-mail).	6 weeks in advance
Anticipate guest count and confirm details with restaurant, including room setup and menu.	2 to 3 weeks in advance
Select baby's attire—something red always delights. The baby wears all the jewelry received during the party.	2 weeks in advance
Compile favors (optional).	1 to 2 weeks in advance
Arrange for decorations (optional).	1 week in advance
Select guest book or traditional Chinese red silk embroidered tablecloth that serves as a guest registry.	1 week in advance
Gather all party items for the party: • *Decorations* • *Favors* • *Guest registry* • *Baby*	Day of the event

big

longevity

On my big birthday, let me treat you to a big dinner. You and you and you. To-day, my son's wife boils sweet egg tea with lotus to prolong my vitality. To-night, I don the long-life robe as my sons honor my years. Neighbors all, come and feast on a life fulfilled with noodles as long as my life, fish as full as the ocean, and duck that shines imperial.

Turning sixty years old is a significant achievement for the Chinese in America. It marks the completion of the Chinese calendar's entire cycle of the twelve Chinese astrological animals as they are associated with the five elements of metal, wood, water, fire, and earth. Considered the age of longevity, a big birthday is life's third milestone to celebrate after birth and marriage. And the celebration is often marked with a multicourse banquet for the honoree's closest family and friends that can multiply into several hundred guests.

Big birthday celebrations known as *dai san yat* in Cantonese begin at age 50, marking a half century, and follow every decade thereafter. Men celebrate their noteworthy birthdays in the even decade years (e.g., 50, 60, 70, 80, 90, 100), while some women choose to celebrate in odd decade years (e.g., 51, 61, 71, 81, 91, 101), depending on the practices of their region of origin. It is also common to see significant birthdays celebrated a year early at age 49, 59, 69, 79, 89, or 99, based on the Cantonese term *cheung so,* meaning "seizing longevity." But regardless of the celebration's timing, the event is always initiated and hosted by the children of the birthday celebrant.

In old China, individual birthdays were often not acknowledged. In fact, annual birthday celebrations were frowned upon until attaining age 50. Instead, birthdays traditionally were celebrated en masse on the seventh day of the Chinese New Year—the day considered Everybody's Birthday—during which everyone automatically advanced one year.

When life span was shorter in old China, living through a full lunar calendar cycle while overcoming life's struggles and strife was considered a great achievement. Furthermore, based on the Confucian principle of filial piety, elders were honored and revered simply for attaining "old" age. Therefore, in a Chinese family, the elders are held in highest esteem. All stand when elders enter a room and offer seats to them. Children are taught to greet them. Elders are served first. They eat first. They depart first. They have the final word.

counting the years

Calculating the age of a Chinese elder requires only a few conditional factors. No need to worry, an abacus is not required. For starters, at birth, a baby already is considered to be one year old, since the Chinese measure time spent in the womb.

When the New Year arrives, everyone becomes a year older during

Everybody's Birthday, the seventh day of the new lunar year. Therefore, a baby who's born on Chinese New Year's Eve could be considered two years old just eight days later.

triple b—big birthday banquet

Planning a big birthday celebration—beginning at the decade birthday of your choice and every ten years thereafter—takes the same organizational skills as those needed for coordinating any large social event. Banquets are elaborate multicourse dinners lasting three to four hours. With some guests traveling from great distances to the celebration and with the presence of family, friends, and business acquaintances, the banquet becomes a reason for reunion.

The steps for planning the big birthday celebration begin at least four to six months in advance with setting a date and securing a restaurant with banquet facilities. The following activities then fall into place like *mah-jong* tiles:

- Printing invitations
- Finalizing the banquet menu and beverage arrangements with the restaurant
- Securing a photographer (optional)
- Arranging for the host table's floral centerpiece(s) and corsages and boutonnieres for family members (optional)
- Retaining a lion dance troupe (optional)
- Planning the evening's program to include speeches and acknowledgments

the banquet menu

Because of the complexity of the food preparation, banquets are usually held in a restaurant, rather than at home. The birthday banquet is several courses and served family style. Depending on the honoree, the birthday celebration could consist just of the immediate family seated at one to three round tables of ten people per table. An alternate option is to hold a more elaborate banquet of several hundred guests. The decision is usually a matter of economics in the end, as the honoree's children customarily assume the dinner tab as well as most incidental costs.

The Cantonese birthday banquet consists of nine courses ("nine" meaning "everlasting"), not including rice or dessert, and always includes long-life noodles to symbolize longevity. Birthday dinners often conclude with a small peach-shaped steamed bun with sweet bean paste called *shau toh* (*shou tao* in Mandarin), which symbolizes the longevity peach. Eating *shau toh* during a birthday banquet is a wish for many, many years of long life and good health.

A typical banquet menu includes the following:

- Appetizer, such as a platter of cold meats and jellyfish, or individual barbecued quails
- Shark's fin or bird's nest soup
- Roasted or braised whole chicken
- Sautéed vegetables, such as black mushrooms with seasonal Chinese greens
- Peking duck with steamed buns
- Honey-glazed prawns and walnuts
- Sautéed beef with Chinese broccoli
- Steamed whole fish in green onion and ginger sauce

- Long-life birthday noodles, fried or steamed rice
- Longevity *shau toh* bun and/or birthday cake

invitations

The traditional invitation of gold lettering on red cards, printed in both Chinese and English, sets the birthday festivities in motion. Chinatown printers have an assortment of styles to select from and will even help with translating names from English into Chinese and vice versa. Be sure to include RSVP cards with postage-paid return envelopes inside the invitations so you can help the banquet chef anticipate how many fish and fowl need to be ordered for the big evening.

long-life robe

In attaining longevity, the birthday honoree has earned the privilege of wearing the traditional long-life robe. Long-life robes mark life's achievement and fulfillment. When the long-life robe is worn, it suggests a person's readiness to obligingly enter the next phase of existence.

For men, the robe is midnight blue with metallic threads woven through the silk fabric, which is often called "treasure blue." A subtle embossed pattern is often visible on the regal garment. Today, however, many men opt to wear tuxedos or a dark tailored suit and coordinating tie. A woman's long-life robe is actually a black silk ensemble consisting of a long skirt and jacket that is ornately embroidered with colorful flowers and birds. Some of the beautiful needlework symbolizes heaven, as these robes are considered funereal garments as well. Dark purple is often woven into the colors, as it's a color associated with death. The last time the robe is donned is typically when life ceases. The thought is bittersweet but necessary for reaching life's ultimate station.

Auntie Lao says wearing the longevity robe is a well-deserved honor and cause for celebration upon death.

banquet flowers

Flowers are selected according to a family's tastes and preferences. The guest of honor and the host family sit at the banquet's head table, which is decorated with one or more floral centerpieces. The birthday honoree and family members wear corsages and boutonnieres so they can be identified easily by the guests. The matriarch of the host family typically wears a regal orchid corsage. Below is a traditional Chinese floral chart for making flower selections by month or season:

Month	Flower
January	Plum Blossom
February	Peach Blossom
March	Peony
April	Cherry Blossom
May	Magnolia
June	Pomegranate
July	Lotus
August	Pear
September	Mallow
October	Chrysanthemum
November	Gardenia
December	Poppy

Season	Flower (continued)
Spring	Tree Peony
Summer	Lotus
Autumn	Chrysanthemum
Winter	Plum Blossom

drink cup! 乾杯

"Yum bui" is a Cantonese toast that literally means "Drink cup." It's the equivalent of saying "Bottoms up." In Mandarin, the universal toast is *"Gan bei,"* which means to drink until the cup is dry. Birthday banquet tables are usually set with a bottle of Chinese liquor, such as the pungent mao tai, or whiskey for toasting the guest of honor. Toasts are conducted from the speaker's podium if using one, and it's polite for all guests to stand and face the honoree.

While dinner is being served, many birthday honorees opt to visit each table with a bottle of liquor to toast their guests in gracious gratitude. For the designated driver or nondrinker, soft drinks, sparkling cider, and hot tea are also available.

While liquor for toasting is a must, wine at banquets is gaining popularity in America. Despite the wide range of dishes served throughout the evening, selecting an appropriate wine does not require the knowledge of a sommelier. Some banquets offer both red and white varieties. But, should only one wine be served with dinner, a flexible, light-bodied white such as a Riesling or Gewürztraminer would please the palate and complement the progression of dishes and varied scale of flavors, textures, and temperatures.

chinese time

Although the banquet invitation slates a specific start time, dinner is frequently served one hour later. This delay has come to be known informally as Chinese time. This is considered a social hour to await the guests' arrival, sign the Chinese red silk guest registry tablecloth, chat and mingle, and locate assigned seating. Some banquets designate seating for every guest, while others opt to reserve only a few tables for close relatives and special guests and leave the remaining tables open for guests to sit wherever they wish.

a few good words

Once the majority of guests have arrived, introductions and honorary speeches officially kick off the banquet festivities. A master of ceremonies, if there is one, will mark the special occasion by first introducing the evening's guest of honor and immediate family members.

Next, introductions usually proceed by birth order. However, traditional families may choose to recognize their sons' families first. Following them are the honoree's brothers and sisters as well as the family members of the honoree's spouse. As a sign of respect and recognition, acknowledgments always include the older generation's aunties and uncles.

benevolent donations

In times of celebration, the Chinese share life's good fortune. During the speeches, the host family may choose to make a series of monetary donations to social organizations, civic groups, and related causes with which the honoree is affiliated. Donations are a gesture of goodwill and gratitude stemming from the idea of reaping what you give. The karmic wheel rewards acts of generosity and is an invitation for further happiness. In the spirit of giving, the Chinese believe what is given ultimately will be returned.

Eventually, donations are individually distributed to the designated entities, and respective representatives officially accept the gifts with words of appreciation for the honoree and the celebratory occasion.

lion dancers

A lion dance, often performed by a local martial arts troupe, is an entertaining highlight of a birthday banquet. Pounding drums and clanging cymbals command the crowd's attention as acrobatic lions wind their way through the banquet room to the honoree's table. The lions please the crowd as they dance, roll, flirt, and bow. Members of the head table reward these bejeweled creatures with a feast of red envelopes for bestowing good wishes and protection. When the lively lions are satiated, they retreat from the banquet room just in time for the serving of the evening's first course.

Birthday Cake

For some Chinese Americans, big birthdays aren't complete without a cake of flickering candles that warrant a small fire extinguisher. At the banquet, a birthday cake is often served according to Western tradition with a gleeful rendition of "Happy Birthday."

birthday favors

No one goes home empty-handed from a Chinese birthday banquet. Besides the take-out boxes of leftovers from the evening's meal, red envelopes of lucky money, or *lai see*, are passed out. Usually containing a dollar bill or coin, the envelopes are given to each guest and financed by the birthday honoree as a blessing in celebration for longevity.

In addition to the *lai see*, birthday celebration guests traditionally receive a rice bowl and a pair of chopsticks to symbolize continued contentment. Many families arrange to personalize the bowl by imprinting their family name in Chinese characters.

Favors for banquet celebrations invite imagination. Today, birthday banquet favors have evolved from the traditional rice bowls and chopsticks to mementos such as statutes of the God of Longevity, teacups, and gold coin tassels. In the West, chocolate truffles or mints packaged in red could serve as souvenirs of thanks.

the god of longevity 壽星公

Shou Xing is the God of Longevity. Always smiling, *Shou Xing* is a gentle, portly, grandfatherly figure with a round bald head and long white beard. Translated, his name means "Star of Longevity." It is said he lives in a palace at the South Pole with a garden of herbs including the herb of immortality. Also referred to as the Old Man of the South Pole, in one hand, *Shou Xing* holds a large jade staff with a dragon's head and in the other, the peach of immortality. He is often depicted with a deer or a crane, symbols of longevity.

birthday gifting—chinese style

Selecting a birthday gift for the person who has achieved long life with nearly all the material possessions that go with it is like pursuing a perfect, meaningful birthday gift for your mom or dad. How do you wrap up wishes of continued health and happiness for an elder in a graceful and symbolic way? Many guests fulfill this tall order by giving red envelopes of lucky money, framed twenty-four-karat-gold Chinese characters, symbols and images called *gum pi* in Cantonese, and decorative works of art.

giving lucky money

Lucky money is the universal Chinese gift for all occasions. It's a highly convenient gift because it only requires a visit to the ATM and a red envelope. When shopping for *lai see* envelopes for big birthday occasions, look for the Chinese character *shou*, meaning "longevity," or an image of the God of Longevity (see page 180).

Currency denominations are given based on your relationship to the honoree, but nonrelated guests attending a banquet may give twenty to twenty-five dollars per adult. If the birthday person is a close relative such as a grandparent, parent, sibling, auntie, uncle, or cousin, the gift may become extravagant. In such instances, family members belonging to the same branch of the family's tree often join together to give a significant gift.

Auspicious dollar amounts include using the lucky number of "eight" to wish continued prosperity, or the number "nine" to bring everlasting happiness and long life. Hence denominations of $88, $99, $888, or $999 are extremely symbolic. Red envelopes are perfectly acceptable gifts and they needn't be wrapped. They are simply handed directly to the birthday honoree or enclosed inside a birthday card. Regardless, whenever giving a *lai see* as a gift, remember to include your name directly on the backside of the red envelope so that your gift can be recognized.

giving a gift of gold

Often during the banquet, all the *gum pi* are displayed on a gift table for guests to view. *Gum pi* are considered traditional and generous gifts. They are twenty-four-karat-gold Chinese characters or symbols mounted on a red background and framed under glass in a gold frame. The characters state congratulatory phrases and special wishes for continued long life, health, prosperity, and good fortune. Many *gum pi* also include symbolic images such as the God of Longevity, a pair of peaches, and the Eight Immortals. *Gum pi* are available in Chinatown's jewelry stores and are sold by weight according to the prevailing price of gold. Depending on the size, weight, design, and your powers of negotiation, *gum pi* usually range from $100 to $300.

Recently, gold statues have grown in popularity as an auspicious gift. Many are in the shapes of the Chinese astrological animals, the Chinese gods, the Eight Immortals, and symbolic flowers such as the peony and lotus. A gold coin is another popular present. When shopping for a woman, a gift of Chinese gold or jade jewelry always warms the heart for celebrating a milestone, any milestone.

Auntie Lao says never to shop for jade or gold on an empty stomach, or you'll go hungry.

giving chinese decorative art

The Chinese believe they invite good fortune into their lives by surrounding themselves with auspicious symbols and phrases. This is a Chinese form of self-actualization. No wonder the decorative arts and crafts, such as paintings, scrolls, porcelain pieces, and statues, are filled with symbolism and remain popular gifts for a big birthday celebration. Ever-present traditional Chinese images and motifs perpetually convey wishes for fortune, prosperity, longevity, happiness, and health.

福 祿 壽 喜 康

Visual puns—called rebuses—are frequently based on many Chinese words that sound similar to one another. Because each Chinese character possesses several tones (nine for Cantonese, and four for Mandarin), many different-meaning words are used simply for their similar-sounding qualities. For example, "fortune" and "bat" (as in Batman) are rebuses because they sound the same (*fu*), though they are written differently. Hence the bat is a symbol of luck and a popular motif. The thought is carried forth in the popular imperial rebus of the *shou* (longevity) character being surrounded by five bats to represent the five blessings of virtue, many sons, wealth, longevity, and a peaceful death. Another Chinese pun is the word for apple, *ping gwa*. Because *ping* sounds like the word for peace, the apple has come to symbolize peace. Other rebus puns include deer (prosperity [*lu*]) and fish (abundance [*yu*]).

The use of rebuses and motifs to convey good wishes for longevity, prosperity, and happiness has resulted in the creation of graphically sophisticated Chinese art forms. Here are some popular images appropriate for a big birthday.

Typical Chinese Symbols for Longevity

- Tortoise
- Pine
- Peach
- Crane
- God of Longevity
- *Shou* character for long life
- Deer
- The Eight Immortals and their accessories (fan, gourd, fish drum, lotus, basket, sword, wood clappers, flute)

Typical Chinese Symbols for Prosperity

- Fish
- Coins
- Gold ingots
- Peony
- Deer

The deer is especially symbolic, as it represents both longevity and prosperity.

食 mom's brown tea-egg soup

serves 10

Several times a year, my mother, Mary, would prepare brown tea-egg soup out of tradition and honor for my grandparents. Tea steeps in the broth for a few hours, which results in an amber-colored soup that also dyes the eggs brown. Mom would wrap a towel around a stockpot of the sweet brown concoction and load it into the backseat of the station wagon for a visit to my grandparents' home, meanwhile instructing us kids to keep a close eye on it. Known as gai don cha in my southern village dialect, this sweet soup contains hard-boiled eggs, lotus seeds, lily bulbs, and dried longan. It's a family favorite that I think of as a special Chinese birthday treat. Brown tea-egg soup is served warm. Remember that the birthday honoree's serving should include a pair of boiled eggs to double the wish for strength and longevity. When serving elders, Auntie Lao reminds us to hold the bowl with two hands when offering it as a gesture of respect.

> ½ **cup dried lotus seeds**
> ½ **cup dried lily bulbs**
> ½ **dried longan**
> **4 to 5 dried persimmons (optional)**
> **5 quarts water**
> **2 Lipton tea bags**
> **7 rectangular Chinese brown sugar bars**
> 1½ **cups granulated sugar**
> **12 large eggs**

1. Rinse the dried lotus seeds, lily bulbs, and longan in separate bowls. Soak in separate bowls overnight.

2. Rinse the dried persimmons if using. In a stockpot, combine the water, lotus seeds, and persimmons. Bring to boil and simmer for 1½ hours. In a separate pot, cook the eggs until hard-boiled. Rinse, let cool, and peel. Set aside.

3. Add the lily bulbs and longan to the stockpot. Insert the tea bags into tea infusers and drop into the stockpot. Add the Chinese brown sugar

bars and the sugar. Simmer for 1 hour. Remove the tea bags. Add the hard-boiled eggs and simmer on lowest heat for 1 hour. Turn the heat off and let sit until time to serve. Serve warm.

big birthday banquet planning guide

If you can anticipate when your father's or mother's auspicious day of longevity is, planning a big birthday celebration falls into place from there. Banquet dates are selected based on practicality, availability, and convenience for the family and their guests. For January and February birthdays, Chinese New Year should be considered. Here are some suggestions for coordinating a big birthday celebration:

Activity	Suggested Timing
Anticipate banquet date and number of guests. Begin compiling invitation list.	6 months in advance
Secure restaurant facilities and establish banquet menu.	5 to 6 months in advance
Secure photographer (optional).	5 to 6 months in advance
Retain lion dance troupe (optional).	5 to 6 months in advance
Select and order favor elements.	2 to 4 months in advance
Order invitations and begin compiling invitation mailing list.	2 to 3 months in advance
Select florist for centerpiece(s), corsages, and boutonnieres.	2 to 3 months in advance

Activity	Suggested Timing
Anticipate banquet decorations (optional).	2 to 3 months in advance
Finalize invitation mailing list and begin addressing invitations.	6 to 8 weeks in advance
Arrange/purchase celebratory attire—long-life robe, new suit or dress. The birthday honoree's outfit should be new from top to bottom.	4 to 6 weeks in advance
Mail invitations.	4 to 5 weeks in advance
Order birthday cake from a favorite bakery and order *shau toh longevity buns* from a dim sum carry-out house. (Preferably, these businesses will deliver to banquet restaurants. Confirm final guest count with them once all RSVPs are received.)	4 weeks in advance
Begin preparing favors.	3 to 4 weeks in advance depending on guest count
Collect RSVP cards—make any necessary calls to confirm attendance.	2 to 3 weeks in advance
Confirm estimated number of guests with the restaurant; make linen selection and provide banquet room configuration for the reception table, host table, gift table, speaker's podium, and any other activities you require for your banquet festivities.	2 weeks in advance
Prepare table tents for seating arrangements.	1 to 2 weeks in advance

Activity	Suggested Timing *(continued)*
Make liquor and additional beverage purchases if necessary.	1 to 2 weeks in advance
Purchase and prepare red envelopes (lai see) for each guest.	1 week in advance
Coordinate, gather, and confirm delivery for all components for the big banquet:	3 days prior to the day of event

Coordinate, gather, and confirm delivery for all components for the big banquet:

- *Banquet room decorations, including reception table signing cloth and black markers*
- *Big birthday favors*
- *Red envelopes containing $1 to be distributed to each guest*
- *Liquor and supplemental soft drinks*
- *Banquet attire*
- *Seating assignment items*
- *Flowers*
- *Birthday cake*
- *Shau toh buns*

burial   *ceremony*

There is no need to mourn

when I am gone,

I live in the quiet

of your memory.

The loss of a loved one is never easy. While emotions run deep, a family is immersed in making decisions and plans for the ritual of life's passage. The Chinese American funeral is further complicated by cultural practices that straddle Eastern rites and Western beliefs. Although the customs may appear incompatible or even contradictory, the deceased and the family share in cross-cultural similarities of solemnity, respect, and memorial of a life well lived.

In old China just prior to death, an elder would be bathed and dressed in a longevity robe for the journey to heaven. At the moment of death, three grains of rice would be placed on top of the decedent's tongue so the soul wouldn't enter the otherworld with an empty mouth. The wealthy would lay a gold piece or a pearl on the deceased's tongue. The Empress Dowager Ci Xi had a beautiful Eastern sea pearl inserted into her mouth so she would go in wealth and not poverty. Then a silver coin would be placed on top of the decedent's mouth as a toll to cross the river of death. As a final act, a white or yellow piece of paper was laid over the deceased's face.

As with funerals everywhere, a traditional Chinese funeral requires planning and organization. Furthermore, during this time of transition, certain measures are also taken to console and protect the living. In the early days, noodles were not eaten until the completion of funeral services, because noodles previously symbolized the decedent's long life. As a sign of mourning, no makeup or jewelry was worn. During the period of mourning, the family wore only either all-black or all-white clothing, the latter also being an Asian color of mourning.

A full Chinese funeral ceremony often consists of an evening wake, family wake dinner, funeral service, processional, cemetery burial, and longevity dinner. In Chinatown, the weekends are popular times to hold funerals to accommodate work schedules for extended families and guests. However, the Chinese calendar can play a significant role in selecting the most auspicious time to assure a safe and successful journey to the spirit world.

> *During Chinese New Year, the Chinese do not perform burials for the first seven days of the year.*

For assistance in deciphering traditional Chinese rituals, funeral homes in Chinese communities are well versed in various Chinese customs, whether they're native to northern China, Shanghai, Taiwan, Guangdong, Taishan, Fujian, Hunan, or from other overseas Chinese communities such as Vietnam and Singapore or other Southeast Asian countries. A benevolent association can further lend some cultural assistance, as can a Chinese auntie who is knowledgeable in the old ways.

Benevolent Family Association

A family belonging to a Chinatown benevolent association will find that it can serve as a source of support and assistance in fulfilling the traditional funeral rites. Chinatown benevolent associations were founded in the 1800s to assist new immigrants with settling in Gum San (meaning "Gold Mountain," the term coined for California during the era when gold was mined). These associations provided a sanctuary for immigrants sharing a common geography, language, and culture. The organizations provided a social network through which immigrants could find shelter, food, and work in a foreign land. Moreover, the associations often handled legal tasks such as resolving business or personal disputes, and arranged funerals and the subsequent shipping of remains back to China.

For many early immigrants, the Benevolent Family Association was a place to gather because it smelled and sounded like the motherland an ocean away. The association clubhouse always offered an open mah jong table, a hot cup of tea, and a chair to rest sore feet. It was a place to relieve a feeling of isolation.

Today, benevolent associations still have an active presence in America's Chinatowns and continue to provide a cultural connection. The organizations hold celebratory banquets and events for various Chi-

*nese holidays, such as Chinese New Year,
Qing Ming (Clear Brightness Festival), and
Chong Yang (Double Ninth Day). Many are
nonprofit organizations that provide support for Chinatown's Chinese language
schools and temples of worship, and many
are also civic and politically active. Often
represented as a unified whole by the Consolidated Chinese Benevolent Association,
also known as Look San, or Six Companies,
the group represents the six major regions of
southern China from where the early wave of
Chinese settlers emigrated.*

funeral flowers

Flowers are the consummate symbol of mourning to the Chinese. It is a common practice for close relatives to send floral wreaths to the funeral home as a sign of respect. The scope of the arrangement often depends on the relationship and acquaintance to the deceased.

Many Chinatown florists specialize in large round or heart-shaped wreaths with wide sashes attached that identify the senders' names and their relationship to the deceased. Often to express their loss, the deceased's adult children will send extravagant floral arrangements in the shapes of a cross, pillow, broken heart, or broken wheel. The decedent's oldest son provides the spray of flowers that sits atop the casket. The color of the flowers can vary according to the age of the deceased. If the decedent was seventy-five years or older, colorful flowers of red or pink are used in recognition of a long life, because death in old age is considered a celebration of life. But if the deceased passed young, only light-colored flowers of white and yellow would be used. For the funeral services, the decedent's portrait is often framed in flowers, where a small imitation bird sits atop to symbolize the flight to heaven.

packing for the otherworld

The funeral home will assume full responsibility for embalming and readying the body for viewing by requesting items such as the longevity robe or a full set of clothes (including stockings), dentures (if any), and a photograph for matching makeup. Many funeral homes will cut the clothing up the back for ease of dressing, but in the Chinese culture, it is imperative that the clothing is never cut. Moreover, in some traditional Asian cultures, pockets are sewn to prevent bad luck from entering through their openings.

For the decedent's journey to the afterlife, the traditional Chinese often provide additional supplies to be packed with the deceased. These items may be three changes of clothing for each realm of existence (heaven, earth, otherworld), three bags of provisions containing grains such as oats or wheat, tea, coins, and fruit to serve as (1) the transportation ferry fee to the underworld, (2) a gift for the guards standing at the gates of the underworld, and (3) gifts for the family's ancestors. Also included are beloved items such as a favored calligraphy brush, a lucky *mah jong* game, or even a bottle of mao tai or cognac. If the decedent's nails are clipped after death, they are gathered into a small bag and also inserted into the casket.

Other items provided for the trip to the otherworld are two passports that include the name of the deceased and date of death. One passport is placed in the hand of the deceased for entry into heaven and the other is burned at the conclusion of the cemetery burial.

security blankets

To keep the spirit warm and help brave the cold, children of the deceased provide several *pei,* or blankets of cotton covers in various colors and prints, to be laid over the body in special order. The oldest son usually supplies the top and second blankets. The first *pei* is solid white, the color of heaven, and is followed by one of solid red, the color of life and happiness. All other blankets are neatly folded so that all the different fabric layers can be seen. Blankets are an eternal gift of warmth and comfort children give a beloved parent for the next life.

If the family wishes, a *shau pei,* or longevity blanket, may be laid over the deceased so that it takes the appearance of a comforter, as its construction is richer and slightly padded. The final blanket shrouding the deceased is the *fut pei,* a blanket of spirituality, which is typically gold with Buddhist prayers printed in red. There is also a Christian version of the *fut pei* that is white and has a large gold cross.

Many funeral homes specializing in Chinese rituals have blankets available for purchase, as do spiritual supply stores.

Auntie Lao says a widow should break her husband's comb and place half of it in his casket and toss the other half on the floor of the chapel or temple. This act breaks her matrimonial bond to her husband and permits her to remarry.

food offering

The Chinese provide food offerings during the funeral services so the new spirit is well nourished for its journey. Lighting incense facilitates the spirit's ease of travel between the realm of the living and the otherworld. The food selections are similar to the offerings for Qing Ming and Chong Yang, but in larger servings, as a new spirit has a long passage through the gates of justice before reaching its ultimate destination.

During the evening wake and funeral service, the food offerings to be fed to the dead include the *saam soong*, or "three dishes": roast suckling pig's head, a whole white chicken, and sautéed vegetables called *jai choy*. The offerings are accompanied with three bowls of rice, three cups of wine, and three cups of tea. A pair of chopsticks stands in each rice bowl to help the spirit eat. Some families may also include fresh fruit, such as bananas, apples, and oranges, on the offering table, as well as *dim sum* items of steamed sponge cake, steamed pork buns, and egg tarts.

paper replicas

To help make the spirit as comfortable in death as in life and to set up a home in the otherworld, miniature replicas of household items are displayed during the funeral services and then burned in an incinerator at the end of the burial. There are a multitude of items to select from: two-story houses in either Western or Eastern style, cars, airplanes, man-and-woman servants, paper clothing, gold bullion stacked in a pyramid, baskets of gold ingots, a television playing Chinese opera, microwaves full of microwavable dishes, rice cookers, refrigerators, cell phones with battery and recharger, boom boxes, desktop computers, window fans, floor heaters, and a washing machine with tubing, agitator, and electrical cord. For the *mah jong* spirit-enthusiast, there is a *mah jong* game complete with tiles, table, and stools. Culinary baskets of cookies, candies, cognac, cigarettes, Chinese black mushrooms, and abalone are available, as are *dim sum* favorites like *ha gow* (shrimp dumplings) and *siu mai* (pork dumplings) in bamboo steamers. For women, there is a jewelry box containing Chinese gold and jade bangles, matching earrings, and a watch.

Lindsey looked at the plastic packages. Each individually wrapped bundle contained a three-dimensional, colored-paper version of an item on the list: a cardboard cell phone, neatly creased fake shirts, cartons of phony cigarettes, a wrist-watch with digital numbers painted on the face, and a puffy pair of paper shoes with a big Nike swoosh. The cashier explained that all these items, in addition to the fake money, were to be burned so that the deceased would have these necessities in the afterlife.

—Kim Wong Keltner,
The Dim Sum of All Things

funeral dress code

In America, the immediate family in mourning typically dresses in all black. All white is also worn as it was in old China and is still now by Chinese Vietnamese and other traditional Chinese cultures. A black waistband along with a black armband denotes the eldest son. Other closely related male family members wear only armbands. Women wear black veils and armbands. Traditionally, men wore armbands on the upper left arm and women wore them on the right arm. However, today both sexes wear armbands only on the right arm. In any case, Auntie Lao will notice.

Grandchildren also wear black armbands. In addition, granddaughters and nieces wear a sprig of evergreen clipped behind their right ears. Great-grandchildren wear armbands with a wisp of green yarn pinned on them to signify the green flower of a descendant. The variance within the black armbands reflects a family's generations.

> **Auntie Lao says that should anything fall to the ground during a funeral, leave it!**

exit and entrance packets

Upon entering and exiting a Chinese wake and funeral service, everyone receives packets to ease the bitterness of loss. Male relatives—nephews, cousins, or grandsons—are assigned the task of handing out small white envelopes, each of which typically contains a piece of cellophane-wrapped hard candy and a nickel, when guests enter the funeral home or chapel. The candy is intended to ease the sorrow and should be eaten either during the service or just upon leaving. When departing, everyone receives a small red envelope containing a quarter meant for buying more sweetness before heading home. When accepting these packets, take them in silence, and continue on your way.

> Auntie Lao says not to return directly home from any type of funeral service. A stop at a convenience store is a good detour for dodging any lingering ghosts you may have picked up.

evening wake

The evening wake for family viewing and last respects typically occurs the night before the funeral services. Traditional Chinese wakes could be Buddhist or Taoist, accompanied by chanting, chiming, and drumming. The service includes the burning of sacrificial items, such as joss paper, otherworld money, and Buddhist Passover money, with prayers for the gods. Flames flutter from red candles while large chamois-colored incense continuously smolders. The swirling smoke becomes a thin veil separating life and death and creates a trail for the spirit to follow from one world to another. The immediate family is usually seated in an alcove at the front of the chapel. Other relatives wearing black armbands enter by gender and birth order. Boys first, girls second.

Traditionally, the evening wake was a private mourning service where candles, joss paper, and incense would burn continuously throughout the night to light the dark path to the underworld. Family members would kneel on straw mats and verbally express their loss. Today, however, paying personal respects by kneeling has been replaced with bowing three times, for the past, present, and otherworld, all done with little verbalization.

Some traditional Chinese families will hire professional wailers to demonstrate to the heavenly gods how sad the family feels about its loss. By expressing how respected and beloved the decedent was in life, the family could encourage the gods to be more compassionate as the new spirit enters the other realm. The women wailers usually sit at the side of the chapel and also perform tasks such as lighting the incense, burning the joss paper and otherworld money, and arranging the food offering.

A rarely practiced old-world Chinese ritual consists of "sin-eaters" and Buddhist priests who chant and exorcise the sins from the deceased into an array of *dim sum* items positioned at the floor of the casket. The sin-eaters consume the symbolic *dim sum* so that the deceased will enter the otherworld free of any wrong. This ritual is not widely practiced today, as securing sin-eaters is cost-prohibitive for a mid-American income.

Following the wake, a family dinner is held at a nearby restaurant. The meal is usually simple and includes *jai choy*, a stir-fried vegetarian dish, and tofu, which is naturally white and symbolic of funerals because of its blandness in taste and color.

funeral service

In America, many Chinese Americans conduct a funeral service with Christian prayers and teachings at a chapel. Paper replicas, lit incense, and food and drink offerings may be displayed from the night before the service while a bilingual Christian cleric officiates with an invocation and a Westernized eulogy. Floral arrangements often crowd the altar and aisles. Relatives, friends, and acquaintances are invited to express their remembrances, followed by words of appreciation from the immediate family. The cleric ends the service with a benediction and conveyance of last respects.

The service concludes when all in attendance file past the casket one final time and the pallbearers transport the casket, flowers, and paper replicas on the last leg of the journey.

funeral procession

Today's funeral processional has evolved from the days when the family carried its loved one's casket while traveling by foot through the rural village, accompanied by a parade of paper replicas, chanting Buddhist priests in flowing robes, noisy cymbals, and incessant drums. In America, the burial grounds often lie on the outskirts of a city at a distance that requires a motorized funeral processional. This entourage of vehicles includes a funeral coach and limousines to transport the immediate family.

In an established Chinatown tradition, a marching band leads the funeral processional while playing classic hymns such as "Amazing Grace" and "Onward, Christian Soldiers." Chinese banners written in characters are carried on foot by relatives. They usually note the name of the deceased and the banner's sponsor, and display declarations such as "Everlasting Longevity and Fortune" and "Five Generations Together" or praises of the deceased's accomplishments. The banner's sponsor typically is a close family member or a community organization in which the deceased was active. An adult son or grandson may ride in an open convert-ible while he holds a life-sized photograph of the deceased so that the community can pay one last tribute. To distract evil spirits and allow an easier passage into the otherworld, yellow strips of devil money, called *kai ching*, are tossed into the street and in front of the family home. Yellow in-cense the length of a Louisville Slugger rides shotgun in the funeral coach to help navigate the way to heaven. As a final farewell, this special tradi-tional procession inches its way through the streets of Chinatown's com-munity, past the places where the deceased once lived and worked.

For a decedent who was a prominent Chinatown leader, multiple bands and several banners could be interspersed between a motorcade of limousines and other vehicles carrying family members. In this larger processional, a motorcycle police escort would be required to maintain order at intersections so that the beginning, middle, and end of the mo-torcade could flow as one.

For decedents who belonged to a Chinatown benevolent association, the procession will stop at the association's community hall along the way. On the sidewalk, an association attendant will have arranged a card table complete with burning incense and food offerings to serve as a rest stop for the new spirit and to mark the spot as the halfway point to heaven. The food offerings are a snack that includes white chicken, steamed white buns, and fresh fruit. At this spot, the children of the de-ceased exit their cars to observe the marching band while it performs a fi-nal hymn before the entire entourage continues on its last leg of the trip.

cemetery burial

The graveside service is often brief. Just a few words are said by the pre-siding cleric. In a Chinese funeral, as the casket is lowered six feet into the earth, each person in attendance will take a handful of soil and coins, which are intended to help the spirit buy its way into heaven, and toss them into the open grave. If individual flowers had previously been given to each attendee, they, too, will be tossed into the grave for a final fragrant farewell.

As the burial service concludes, funerary accessories—black armbands, waistbands, veils, evergreen leaves, even pallbearers' gloves—are removed and dropped into the grave. Every item associated with the funeral stays with the deceased. Nothing is taken away except the red exit envelopes, their contents to be spent on the way home.

As a final token of appreciation for the attendees' presence at the funeral and burial services, the family will often arrange to place a *lai see*, enclosed with five to ten dollars, on the windshield of each person's car.

Auntie Lao says:

- *On rainy funeral days, place a small gong on the deceased's body to protect it from thunder and lightning.*
- *Tears on a casket are like tears on silk; they slow your beloved's departure to heaven.*
- *No looking at the casket being lowered into the grave, or the spirit might catch you.*
- *No expectant parents should be at the funeral or cemetery. There's no mixing of happy with sad, good with bad, living with dead.*

longevity dinner

After completing all aspects of the services, a family gathering for a *shau chaan*, or longevity dinner, is held at a restaurant. Although a series of courses is served, dinner is family style and usually not as elaborate as other milestone banquets. But, despite the number of guests, the number of tables is always odd in order to satisfy the gods with a yang numeral. A sample menu is below:

- Soup
- Roast suckling pig—the succulent meat of the heavens
- Boiled white chicken
- *Jai choy*—sautéed vegetarian dish that's considered the food of heaven
- Tofu—whiteness of which is associated with the other world
- Fish
- Steamed rice

The longevity dinner is to express gratitude to those who paid respects to the departed at the funeral. The dinner also provides an opportunity to relieve the tension and grief accumulated throughout the days leading up to the funeral. The dinner closes a chapter for the family and begins the healing process.

care packages

Care packages are prepared for the immediate family members and are distributed to them before leaving for home. Determining who receives what items depends on the members' various relationships. The care package items are typically assembled in brown paper bags and distributed as follows:

Relative	Care Package Contents
Immediate Family Members—wife, married siblings, married children, married grandsons and granddaughters (single adults are considered their parents' children, thus do not usually receive personal care packages)	• Knife—a symbol for severing the pain of loss • Flashlight (with batteries)—to light the way out of darkness • Juniper—a symbol of evergreen • Chinese brown sugar blocks—to sweeten the bitterness
Married Nieces and Nephews	• Red envelope • Juniper • Brown sugar

When returning home after attending the funeral of an immediate family member, be sure to lay a flashlight in the doorway, pointing it toward the interior of the house. Then step over the light. This act signifies lighting the family's way out of darkness.

gateway to the otherworld

Upon death, the traditional Chinese believe, the soul travels through several gateways of justice, where a life is examined and judged, before reaching its final destination and designation for reincarnation. At these junctures, a tally of all rights and wrongs in life is evaluated and reflections from the mirror of life are assessed. Wrongdoers are forced to acknowledge their evil ways. Souls who lived good, generous, positive, and honest lives may move freely through the gates. It usually takes forty-nine days for the soul to pass through life's reflection and retribution. After this period, the memory slate is wiped clean and the parameters for reincarnation either as a superhuman or a gnat are determined and ordered.

family's first cemetery visit

Either three, seven, or thirty days following burial, the immediate family (spouse, children, and their families) will visit the cemetery for their first *beng san*—the "walk mountain" ritual for providing spirit offerings. Your wise auntie will advise when the mandated visit day is based on your family's regional practices. The ritual is similar to the ancestral one performed during Qing Ming and Chong Yang. But, as this is the family's first ritual for the newly deceased, the food items are more plentiful to better establish the new spirit in its new surroundings. Below are lists of foods and items to bring:

Food Items

1 large piece of roast suckling pig	1 hard-boiled egg, cut in half
1 whole roast duck	
1 whole boiled white chicken	1 orange
2 small whole dried fishes	1 banana
4 white steamed pork buns	1 apple
1 slice of Chinese sponge cake	3 cups of Chinese wine
2 small mounds of steamed rice	3 cups of tea

Spirit Items

One bunch of incense

Two red candles

Joss paper

Paper money

Other Items

Flower bouquets and containers

Matches or lighter

Two metal containers—one for burning joss papers and the other filled with enough kitty litter or other fireproof granules to hold upright the lit incense and candles

The ritualistic practice for nourishing the spirit is:

1. Display all food items on the grave site.

2. Light the red candles.

3. Burn the joss paper and paper money completely down to ash. To facilitate burning, you may fold the square joss paper into an ingot shape by first rolling it into a tube and then tucking in the bottom ends of the tube to create a boat-shaped "gold ingot."

4. Light the entire bunch of incense at once and distribute three incense sticks to each person. Stand the remaining lit incense in the metal container.

5. Each person bows three times as a sign of respect while holding the three incense sticks with both hands and then stands the incense upright in the metal container.

6. When everyone has finished paying respects, the deed is complete and it is time to depart. It's up to the family whether the food offerings are taken home or left at the site.

mourning time

Auntie Lao says that after a family member's death, the surviving family will remain home for forty-nine days before venturing out for social visits or events. Some traditional families remain home for one hundred days. This mourning period is practiced out of Confucian filial piety and it's a time when the family will reschedule special occasions and will withdraw from celebrating the annual Chinese festivals. During Chinese New Year, instead of hanging the gold *fu (fook)* character on red paper around the home, a black character on white paper is hung.

In old China, a family observed a mourning period of three years for the passing of a parent by dressing very simply only in dark blue, black, or white. The old rules also stipulated no wearing of jewelry, no makeup, no partying, and no hairstyling. With the passing of a grandparent, a mourning period of one year was observed.

chinese funeral arrangements and planning

Deciding what to do is never easy after receiving word of a loved one's passing. Here are some steps to help plan and arrange a Chinese funeral with some traditional customs:

Activity	Timing
Select a funeral home that's convenient and respected in your community.	Immediately, if not already established
Arrange for a cemetery resting place.	Immediately, if not already established
Provide funeral home with death certificate, complete set of clothing or longevity robe, dentures (if any), and current photo. If preferred, provide extra sets of pants or clothing, three bags of provisions per the Chinese tradition, and any other personal items.	Within a few days
Establish dates and times for services: the evening wake, funeral services, cemetery burial.	Within a few days
Secure presiding cleric or officiant for funeral services.	Within a few days
Select casket and blankets.	Within a few days
Select and contact pallbearers and attendants to help facilitate funeral activity.	Within a few days
Prepare funeral program for distribution at funeral service.	When time and program is in place
If preferred, seek assistance from funeral home regarding wailers, marching bands, and number of limousines.	Within 2 to 3 days
Arrange for family funeral flowers to be sent to funeral home.	Within 2 to 3 days
Secure restaurant reservations and arrange for dinner menus	Once wake and funeral service dates and times are established

Activity	Timing *(continued)*
for family evening wake dinner and farewell dinner following funeral.	
Purchase paper replicas, small white envelopes for entering, small red envelopes for exiting, and packages of sacrificial incense, joss paper, and otherworld money from spiritual supply store or funeral home.	Within 2 to 3 days
Purchase hard candies and coins needed for entrance and exit envelopes and prepare envelopes.	Within 2 to 3 days
Estimate number of red envelopes needed (one for each car at burial service); enclose $5 to $10 inside for each envelope.	A day before the funeral services
Prepare red envelopes with $20 for each attendant and helper.	A day before the funeral services
Prepare family care packages for returning home. Distribute them at the farewell dinner.	A day before the funeral services
Select black clothing.	The day of the evening wake
For evening wake—distribute entrance and exit packages. Have attendants pass out black armbands to family members and collect them before leaving for the family dinner.	During evening wake
Distribute entrance and exit packages. Pass out to family members black armbands and accessories, which will be tossed at the burial service.	During cemetery service

Activity	Timing
Provide attendants with red envelopes to attach to windshields of cars attending cemetery service. Seek assistance from funeral home attendants with this task.	During cemetery service
Distribute attendant red envelopes to funeral home attendants following the burial service.	Following cemetery service
Distribute other attendant, pallbearer, and helper red envelopes. Distribute care packages to closely related family members.	During longevity dinner

eaten

rice

yet

Have you eaten, yet?

A common Chinese greeting is, "Have you eaten, yet?" It's an inquiry into the state of one's well-being and contentment. The polite response is, "Yes," even if you haven't eaten in the past twelve hours and are feeling hypoglycemic. The greeting is extended in kindness only and not meant as a literal inquiry. This greeting arises from when food was not always plentiful and one's well-being was equated with having eaten a meal.

Dining in a traditional setting with the Chinese does not require an etiquette lesson. However, being familiar with a few cultural nuances will build confidence and provide an opportunity to impress your companions, whether at a grand banquet ballroom or the home of a Chinese friend.

chinese hostess gifts

Gift-giving is a customary formality and an expression of appreciation for a dining invitation. The Chinese are generous and will rarely enter a friend's home—especially for the first time—empty-handed. It's a sign of respect, acceptance of friendship, and thanks for the privilege of being invited. Gift-giving is a way to maintain "face." It allows the guest to reciprocate the invitation with an advance token. The hostess will usually remark that a gift was not necessary and that the guest shouldn't have gone through the trouble for just a "small, simple" meal. But, when this comment is translated into Chinese thought, it's a compliment, because the action of your generosity and effort will leave a lasting impression.

The appropriate gift item often depends on your relationship to the Chinese host, hostess, and family. When visiting a relative's home, food items are appropriate, such as fresh seasonal fruit (oranges, apples, Asian pears, persimmons), candy, cookies, Chinese preserved fruit, and cured beef jerky. If visiting during the day, *dim sum*, pastries, and buns are also appropriate. Chinese elders would especially appreciate luxury items such as dried black mushrooms, dried scallops, shark's fin, bird's nest, Chinese sausage, canned abalone, tea, a bottle of rice wine, or even mao tai. When compiling the gift package, select an odd, or yang, number of items, as it relates to the living. Eight is also a good number because it connotes prosperity. Never give four items; the number four is bad luck because it sounds like the word for death.

Gifts for business and social acquaintances are more formal. Appropriate items could reflect the hosting family's interests and hobbies, such as sports memorabilia, illustrated or pictorial coffee-table books, decorative items for the home, a bottle of fine whiskey or cognac, gourmet chocolates, or other luxury items. If the hosts have children, popular choices are toys, games, and play clothes with contemporary cartoon characters.

Always present wrapped gifts to your host. Consider wrapping in colors fortuitous to the Chinese: red, gold, yellow, or pink. Avoid wrapping in white or black, as they are associated with funerals. Also reconsider green (the color of separation) and blue (the color of mourning).

Never give a clock because the Chinese associate it with death: the word for clock, *jung,* sounds like the Cantonese word for funeral. Watches, on the other hand, are popular and contemporary accessories, as well as being functional. Knives or scissors are inappropriate gifts, especially for business associates, as these items represent severing ties. The Taiwanese don't give umbrellas because the word for umbrella sounds like the word for separation. Handkerchiefs and white flowers are off-limits because they are also associated with funerals. Nowadays, cut flowers are becoming more acceptable as a hostess gift because of relaxed belief with the old superstitions. But little ladies like Auntie Lao would still cringe at the sight of them. When in doubt, give a living potted plant.

> *Always present and accept a gift with two hands as a sign of respect, reverence, and sincerity.*

It is not customary for the Chinese to open gifts in front of the giver. They will graciously accept the gift and put it aside for opening after the guests have departed. The purpose of this custom is twofold. First, the Chinese consider impatience and selfishness taboo. Second, it spares embarrassment to the giver and receiver should the gift not be pleasing, thus "saving face" for all involved. Moreover, the Chinese typically do not send thank-you notes, although many have adopted this gesture in America, depending on the situation in which a gift was bestowed. Because the gratitude has been expressed in person at the time of receiving the gift, the Chinese feel another written formality is redundant.

dining in the home 食

An invitation to dine in a Chinese home is a great honor. It's the ultimate compliment to a friendship, as many Chinese are culturally reserved about the intimacy and privacy of their homes and instead prefer to entertain in a restaurant. This custom arises from the days when the typical home was too small to accommodate any but the family's residents.

Upon entering a Chinese home, if you spot a shoe collection at the entryway, you should also remove your shoes. Follow your Chinese host's lead as to where to sit for the evening. Don't wander through the house even when offered a tour of the home. When touring, always follow and always allow the oldest members to enter a room first out of respect.

When dinner is served, again, take your host's lead for when to begin drinking and eating. Often, a toast of friendship and appreciation for the evening will be given at the meal's start. When dining in a home, respectfully serving the elders first still applies—as does serving the guest seated next to you—until they politely excuse you from the gesture. In any case, serving yourself should always come last.

For a pleasant dining experience, keep dinner conversation to light social topics. Save world politics, the state of the nation, or the evolving economy for another time. Conversation should gravitate to the meal's tastiness and the host's efforts for preparing the dishes. At the meal's end, it's acceptable to the Chinese to use a toothpick discreetly at the table by moving the pick with one hand while covering the entire mouth with the other. When it's time to clear the table, it's polite to offer to help but highly unlikely you'll be taken up on it.

restaurant dining 餐

Whether in China or North America, dining in a restaurant is a universal method the Chinese use to celebrate and entertain, regardless if the occasion is an educational or professional achievement, visiting friends, or a business venture. Upon arrival, the Chinese host will take charge and lead you through an enjoyable and satisfying evening.

The traditional Chinese dinner table is round to signify the unity of the family and often holds eight to ten persons. Seating arrangements follow a general system, although there are several variations. In one style, the two most important seats at the table oppose each other: One has its back closest to the door, and the other is directly across, facing the door. The honored guest will usually sit at the latter position, while the host will sit facing the guest at the former position. To the host's right will be the spouse, followed in counterclockwise fashion by other family members based on age and rank. Strangely enough, this arrangement often leaves the youngest member of the family, or the lowest person in the hierarchy, sitting next to the host and honored guest. When this happens, seating adjustments may be required.

If the table is rectangular, the host sits to the left of the honored guest and both take the two center seats of the long side of the table facing the door, subsequently followed by family members or business associates according to hierarchy around the table.

For an informal occasion, a Chinese place setting includes a beverage glass, teacup, condiment dish, rice bowl, small plate, soupspoon, chopsticks, and napkin. Soup bowls are often placed near in the center of the table for easy serving. Chopstick rests may also be used but are usually saved for formal occasions. Beverages and communal teapots are usually placed on the table when everyone is seated. Here, as always, the elders' glasses and teacups are filled first.

When dining formally, Chinese etiquette dictates not to fill your own glass or teacup, but to attend only to your neighbor's drinking whims. In return, your neighbor will assure you will not go thirsty. Having to refill your own cup implies that your neighbor was inattentive, thus you would have let him "lose face." Thankfully, today in America, the rules are relaxed and more forgiving, so refilling your own glass has become more acceptable. Nevertheless, when refilling for yourself, still check if anyone else's glass needs filling. If the teapot runs dry, placing the lid slightly askew will alert the waitstaff for replenishing. In the traditional drinking style, tea is enjoyed before or after, not during, dinner.

chinese chopsticks

The Chinese invented chopsticks as an eating utensil over five thousand years ago. Ever since, their popularity has continued to grow. A pair of Chinese chopsticks is usually rounded and more blunt-tipped, less decorated, and longer than the Japanese counterpart. One reason for the difference is that the Japanese dine on numerous dishes served in individual containers placed closer to the diner. Conversely, the Chinese typically dine family style, in which everyone shares from platters placed within arm's reach at the table center. Another reason is that the Japanese long ago distinguished their utensil into an art form to be enjoyed for its beauty as well as its utility. The Chinese, however, value the chopsticks' utilitarian nature first.

Chopsticks are created from many materials and made into various shapes. Common materials are bamboo, wood, plastic, and bone. Pre–worldwide ban, ivory was also popular. Fancier chopsticks are made of gold, silver, part-cloisonné, or jade. Many tops are carved with dragons, phoenixes, unicorns, lions, or Chinese zodiac animals. In old China, emperors preferred silver chopsticks because they were said to turn black when they encountered poison. Luckily for the rulers, this theory usually went untested, as royalty had official food tasters. For tasters, many were not as lucky—with or without silver chopsticks!

When using a set of chopsticks, Auntie Lao says:

- Keep the ends of the chopsticks parallel and even in length.

- Don't use chopsticks as impromptu drumsticks, batons, or pointing devices at the dinner table.

- Don't use chopsticks for waving or for directing foot traffic in the restaurant.

- During breaks, rest the chopsticks' eating end on the edge of the plate or on a chopstick rest—never on the table.

- Don't use chopsticks to spear food—only small children are allowed this infraction.

- Reserve chopsticks for picking up food only.

- Never stand chopsticks upright in a bowl of rice, because they resemble the lit incense sticks at an altar table.

- Don't tap an empty bowl with chopsticks—that's reserved for beggars.

- Use chopsticks in tandem with fingers to steady certain foods; use fingers only as a last resort.

- Don't use chopsticks as hair decoration. The Chinese don't put eating utensils anywhere but on the table.

Auntie Lao says the higher a maiden holds her chopsticks in her hand, the farther away she'll move when she marries.

soup's up

Many Chinese restaurants offer multicourse dinners for a table of eight to ten. For this dining, the menu will often be predetermined and written in Chinese on a special board that may also include additional house specialties. This type of dinner is considered elaborate dining—but not at elaborate cost when compared to ordering the same items à la carte.

When deciding on the amount of food needed, a good rule of thumb is to select one appetizer, one soup, plus the same number of dishes as the number of diners. This system will allow everyone to enjoy a greater variety of dishes, each containing different ingredients cooked in a different way, with a wide range of sauces, textures, and colors. Keep in mind that the soup course often sets the stage for how simple or elaborate the rest of the dinner will be. Watercress soup is a perfect opener for less elaborate fare, while winter melon soup or a soup with seafood could start a special family dinner. An elaborate celebratory banquet could begin with shark's fin or bird's nest soup.

When the first course arrives, restrain your chopsticks unless you're the oldest member at the table or the host has picked up his. If the dish has been placed on a rotating lazy Susan at the center of the table to facilitate reach, you should spin the platter to the eldest or to the host to start the serving. If the host defers, it would be even more courteous for the person sitting next to the elder to take the initiative of serving a hearty portion onto the host's or elder's plate. The lazy Susan then rotates to the next eldest until reaching the youngest member. Husbands often serve their wives, and grown children serve their parents. To retrieve food from the communal dishes, it's considerate to use either a serving spoon or chopsticks you have inverted so that the eating ends do not touch shared foods.

If you don't want another helping of a dish, leave some of the previous serving on your plate. This will be a cue that you are not ready for another heaping spoonful of the same. However, if the host or hostess insists on giving you more food, you should courteously accept, then leave it uneaten if absolutely necessary.

Rice bowls should be used for the rice. Plates are used as an intermediary station for individual servings of *soong* or entrée. This is the proper method for eating from dishes on a lazy Susan: (1) Take a helping from the communal dish and place it on your plate; (2) pick up a chopsticksful of food from the plate and place the food in the rice bowl; (3) place the bowl under your chin or directly against your lower lip; (4) lift or push food and rice into your mouth. During the course of the meal, the rice bowl is held in one hand, while the chopsticks occupy the other. Be warned that it's considered barbaric to take food from the serving platter and put it directly into your mouth.

finger tapping

In the southern China regions of Guangdong Province and Hong Kong, there is an old custom of tapping the table to express thanks to the waitstaff. Tapping while the staff pours tea can earn their diligent attention throughout the course of your meal. The gesture is done first by gently rapping your index and middle fingers on the tabletop to show appreciation, then by curling the same two fingers downward to signify humility. This practice stems from an early emperor who traveled incognito with his entourage through lands far from the royal Forbidden City. When he stopped in a restaurant for a meal, he needed to protect his identity and be served in full discretion. Recognizing the situation, the royal minister created the custom of coded finger tapping to imitate the bow and kowtow of appreciation and humility.

variations on chinese cuisine

Have you ever been curious about the difference between Mandarin and Cantonese cuisines? Where did orange chicken and Peking duck originate? How do Sichuan (Szechuan) dishes compare against Hunanese or Shanghainese? After a while, these regional styles sound like a hodgepodge of chop suey—which, by the way, was invented in America, as was the revered fortune cookie. Many Chinese restaurants offer a number of regional cuisines on the same menu, so it's not that easy to distinguish the difference when all dishes are equally delicious. Here's a quick primer on the most prevalent Chinese cuisines found in America:

canton cuisine

Cantonese cuisine originated from the areas of Guangdong Province and Hong Kong in southern China. Canton is an old port city that today is referred to as Guangzhou. *Dim sum,* meaning "touch the heart," the Chinese meal of small tidbits of food presented on roving carts, began in this region. Freshness is supreme to the Cantonese. Live fish and seafood are held in tanks just before being dispatched immediately for cooking. Cantonese sauces are mild and subtle so as not to overpower the freshness of the ingredients. Popular Cantonese dishes include steamed whole fish, crispy-skinned chicken, shark's fin soup, and roast suckling pig.

mandarin cuisine

Mandarin cuisine is the food of the northern imperial courts of old Peking, known today as Beijing. In this region, wheat instead of rice is widely used, as is a pale leafy cabbage, known as Napa cabbage in America. The crepelike wraparound mu-shu pork and crispy Peking duck accompanied with steamed buns originated in this area. Mandarin cuisine, an elaborate style arising from the imperial days, is often intricately decorated with vegetables carved into flowers, animals, and designs. In another northern dish, Mongolian hot pot, diners cook their own meats and vegetables in a large boiling pot of flavorful broth at the table. Other popular Mandarin foods are pan-fried pot stickers, garlic and scallion Mongolian beef, and beggar's chicken.

shanghai cuisine

The Shanghainese have mastered the arts of braising and stewing so full-bodied flavors commingle on the tongue. Generally considered the cuisine of China's southeastern region of Zhejiang Province, the sauces tend to be rich due to slow-cooking techniques and reduction of sauces. The area is also known for preserving food by pickling vegetables and curing meats. Noodle products are heartier as in Shanghai noodles. The region's sherry-colored wine, Shao Xing, is exported worldwide and is an important ingredient in many dishes. Popular regional dishes are cold appetizer dishes such as drunken shrimp and wine chicken, stewed "lion's head" meatballs, sea cucumber with shrimp roe, and pickled greens with pork.

sichuan cuisine

Chili peppers and red peppercorns are used in Sichuan (Szechuan in Cantonese) cooking to stimulate the taste buds and counter the bitter cold of winter. Sichuan dishes are considered spicy; although the heat is not immediate, it can creep up on you. Through pickling and salt-curing, the vegetables and meats of this province are preserved to last through the harsh winter. The combined flavors of vinegar with sweetly fried food originated in this central western region. Well-known Sichuan dishes are Szechuan beef, stir-fried green beans, cold noodles with peanut sauce, and spicy stir-fried ma po tofu.

hunan cuisine

The food from Hunan is hot, hot, hot. It's often difficult to distinguish Hunan from Sichuan cuisine, as many Chinese restaurants in North America tend to serve both regional styles side by side. The cuisines dovetail nicely as the two provinces are also neighbors in China's heartland. The Hunanese use preserved basics such as hearty oils, garlic, and chili-based sauces. The stir-fried meats are often seared prior to stir-frying, creating sauces and dishes that exude comfort. Popular dishes from Hunan are orange beef or chicken, spicy eggplant in garlic sauce, and hot crispy fish.

chinese alcoholic beverages

Whether dining at home, the neighborhood Chinese family–owned restaurant, or the grand banquet hall, there are a host of Chinese beverages to lift your spirits while dining on an array of dishes of varying temperatures and textures. Many Chinese wines are fermented from rice and other grains and tend to be light in color. Here are some of the most popular spirits:

- Mao Tai—made from wheat and millet. This potent liquor is used for toasting in traditional Chinese banquets and celebrations.

- Sancheng Chiew—a common rice wine that is also used for cooking chicken-wine soup for new mothers and other dishes.

- Kao Liang—a clear distilled spirit that is similar to vodka, yet smoother, from northern China.

- Shao Xing—rice wine similar in color to sherry that originates from Zhejiang Province and is used in preparing various regional dishes. It is often served warm like sake.

- Ng Ka Py—a dark yellow spirit from Guangzhou that is similar to strong bourbon and recognized by its stout, dark, flared-neck jug. Infused with herbs, it has a medicinal taste.

- Mui Kwe Lu—a strong liquor distilled from grains that contains the essence of rose petals.

- Cognac, brandy, and whiskey—popular Western spirits that are commonly served during the course of an evening of celebration and special occasions.

- Chinese beer—Chinese lager, such as Tsingtao, is a refreshing accompaniment to a wide variety of foods from spicy General Tsao's chicken to crispy Peking duck to salt-and-pepper shrimp.

chinese tea

What do Water Nymph, Eyebrows of Longevity, and Gunpowder have in common? All are names of the ubiquitous Chinese beverage *cha*—or tea. When visiting someone's home, the host will graciously offer a cup of tea as a gesture of hospitality and a chance to readjust the internal time clock. Tea isn't taken to quench thirst or provide a morning rush, but to awaken and nurture the spirit.

For the Chinese, tea represents an art form. Since the fifth century C.E., much has been written about the essence, production, and ceremony of tea. Two popular legends explain how tea drinking came to be. The first occurred in 2737 B.C.E., when Emperor Shen Nong, divine father of agriculture and herbal medicine, decreed that only boiled water should be consumed after he observed that it improved his people's health. One day while the emperor was in the forest collecting medicinal plants, he asked a servant to build a fire and boil a large pot of water. When the servant tossed more wood into the fire, however, some withered tea leaves fell into the hot water, instantly creating a fragrant scent. The aromatic amber liquid enticed the emperor to a taste, and soon thereafter the boiling and preparation processes were refined until an invention for the entire world was made.

The second story for tea drinking involves the Indian Buddhist monk, Bodhidharma, who brought Zen Buddhism to China in 520 C.E., and is said to have founded Shaolin martial arts. As a sacred teacher of meditation, the monk vowed to sit for nine straight years at the wall of a cave outside Nanjing that is now known as the Shaolin Temple in Henan Province. With every passing day, the spiritual master's task grew more difficult as his lids became increasingly heavy. So, to keep from falling asleep, he took extreme preventive measures by cutting off his eyelids and tossing them aside. It's said that wherever his lids fell, a tea plant sprouted to hinder further sleep.

tea varieties

Chinese teas generally come in six primary varieties: green, oolong, black, pu-erh, white, and flavored. All varieties come from the same plant—the *Camellia sinensis*—yet their tastes differ due to varying techniques of processing and fermentation. Tea is available in many forms, including loose leaf, powder, bags, bricks, and flat round cakes. A brief description of each tea variety follows:

1. Green Tea: Unfermented for a fresh, mild taste. Leaves are picked as shoots or while they are young, small, and tender. The leaves are immediately dried to prevent oxidation. The leaves of green teas are handled as little as possible, allowing them to hold their virginal green color. Green tea is steeped uncovered to prevent overheating. If the tea is too strong or bitter, reduce the water temperature, as boiling water can scald the leaves. Popular green teas are Long Jing (Dragon Well), the tea of Hangzhou in Zhe-

jiang Province that is used for the famous dish tea-smoked duck; Biluochun of Jiangsu Province that was a favorite of Chairman Mao; Zhucha (Pearl Tea), also referred to as "Gunpowder" because it's tightly rolled into round pellets; and Huangshan Maofeng, famous from Anhui Province.

2. Oolong Tea: Semi-fermented tea that's less astringent than green tea but milder than black. Meaning "black dragon," oolong tea is made by picking the leaves at the peak of maturity and rolling them by hand or in a rotating drum until a red edge appears while the interior of the leaves and the stem remain green. Oolong *(wu long)* originated in Fujian Province, where it's widely produced as well as in other provinces on China's southeast coast. Taiwan also produces oolong and is well known for its Formosa Oolong. Tieguanyin (Iron Goddess of Mercy) is a premium oolong that was once harvested by trained monkeys on the steep cliffs of northern Fujian because these tea plants grew too high for men to climb. Oolong provides a wide range of styles, from a light floral green to a rich luscious dark grade. Often served in Chinese restaurants in North America and in southern China, an oolong tea can be considered a staple.

3. Black Tea: Fully oxidized tea for a strong and hearty flavor. Black tea is referred to as "red tea," or *"hong cha,"* by the Chinese. The leaves are set out on flat trays in rooms where temperatures are graduated to wither or steam to serve as a firing process. Once cured, the leaves are hand-curled and twisted, turning them black. When dry, they are packaged loose or in brick form. Popular Chinese black teas are Keemun, a high-grade type from Anhui Province that serves as a base for scented blends, and Lapsang Souchong, a smoky black tea from Fujian Province that's cured over pine fires.

4. Pu-erh Tea: An exotic, earthy dark tea classified in between black and fully fermented. Also known as *po nay* or *bo lay*, it has a distinctive flavor and fragrance. Pu-erh comes from Yunnan Province and is traditionally bought in flat round cakes or brick form that serves as a luxurious hostess gift. It's also available in bags and loose leaf. This is Auntie Lao's favorite tea for *dim sum* because it helps break down the fat in rich food and aids digestion.

5. White Tea: A light and smooth alternative to the three commercial categories of green, oolong, and black Chinese teas. This rare tea is sun-dried and regarded for its health benefits. Well-known white teas are White Peony, also referred to as Eyes of Longevity because of its furry leaf, and Silver Needle.

6. Flavored Tea: Scented tea enriched by the fragrance of flowers or fruit juices. The most famous tea in this category is the fragrant jasmine that has been a favorite since the southern Song dynasty. Other floral-scented blends include teas with the essence of rose, chrysanthemum, orchid, orange blossom, and camellia. A popular fruit-infused tea is Lychee Black from Guangdong Province. This black tea is treated with the juice of the lychee, one of southern China's most cherished summer fruits.

tea infusion

There are two ways the Chinese usually prepare tea: in porcelain teapots or individual lidded cups. The Chinese tea ceremony, although less elaborate than the Japanese, nevertheless is ritualistic by heightening the senses. The accoutrements used are a Yixing clay teapot and small teacups; a kettle, heating stove, and draining tray; and premium tea leaves and water. In the Chinese ceremony, the tea-making and serving steps are just as important as savoring the taste of the tea. Below outlines the procedure for a traditional Chinese tea service:

1. Heat the water to the desired temperature based on the tea leaf variety being served. Lighter teas will take lower water temperatures than darker oolong and black teas.

2. Douse the teapot and cups with hot water to sanitize and warm them before steeping.

3. Add one to two teaspoons tea leaves to the teapot. Pour the heated water over the tea leaves, then pour out the water. In removing the initial infusion, the Chinese feel the leaves are ready for the second—and best—infusion that releases the tea's truest bouquet.

4. Infuse by filling the teapot and covering it with the lid for three to five minutes. Steeping time varies depending on the tea variety. Lighter green teas and oolongs require less steeping than darker oolongs and black teas. High-grade tea leaves will retain their flavor through at least a third infusion.

5. Store tea in an airtight metal or glass canister (not plastic) in a cool, dry place. Tea leaves have delicate oils that impart their flavor and bouquet, both of which can easily change and evaporate.

glossary

In *Good Luck Life*, the Pinyin/Mandarin romanization system has been adopted for formal Chinese names of provinces, dynasties, and festivals. This is the official system for standardizing Chinese that is recognized by the People's Republic of China and the United Nations. For common references, Cantonese words are used that have been phonetically romanized by the author, as there is no official form of romanization for Cantonese. The glossary that follows cross-references between Pinyin and Cantonese. In Pinyin, the critical consonants are pronounced as follows:

1. "C" is pronounced "ts," as in "bats."

2. "Zh" is pronounced "j," as in "jar."

3. "Q" is pronounced "ch," as in "chess."

4. "X" is pronounced "sh," as in "should."

Chapter/ Page	Cantonese	Pinyin/ Mandarin	English
	Ma Ma	Nai Nai	Paternal Grandmother.
1-4	Po Po	Po Po	Maternal Grandmother.
	Yeh Yeh	Ye Ye	Paternal Grandfather.
1-4	Gung Gung	Gong Gong	Maternal Grandfather.
1-8	Maai-lan, maai-lan! Maai-doh-nien-saam-shap-maan!	—	Old Cantonese rhyme of Chinese New Year meaning "Selling laziness, selling laziness! Laziness for sale until New Year's Eve."
1-9	Fook	Fu	Good fortune or blessings.
1-10	Tong dynasty	Tang dynasty	Dynastic rule between 618 and 907 C.E.
1-10	Chan Shook Bao	Qin Shu Bao	One of Emperor Taizong's generals that stands as one of two Door Gods with Wei Chi Jingde.
1-10	Wai Chi Gong	Wei Chi Jingde	One of Emperor Taizong's generals that stands as one of two Door Gods with Qin Shu Bao.
1-10	Aiiya!	Aiya!	Chinese interjection exclaimed during a moment of surprise, such as wow, uh-oh, or oh, no!
1-13	Nien	Nian	The evil spirit or demon. Also, the Chinese word for year.

Chapter/ Page	Cantonese	Pinyin/ Mandarin	English
1-13	Pau jeun	Bao zhu	Firecrackers.
1-15	Gung Hay Fat Choy	Gong Xi Fa Cai	Wishing you happiness and prosperity.
1-15	Sun Nien Fai Lok	Xin Nian Kuai Le	Happy New Year.
1-15	Lai see	Hong bao	Red envelope of lucky money.
1-15	Doi jeh	Duo xie or xie xie	Thank you.
1-15	Hoong	Hong	The color red.
1-16	Jai choy	Zhai cai	"Monk's" vegetarian dish containing many auspicious ingredients that is eaten at Chinese New Year.
1-17	Fat choy	Fa cai	Sea moss, a food ingredient that looks like long black hair.
1-17	Fun see	Fen si	Long, clear bean threads.
1-17	Gai lon	Jie lan	Long-stemmed leafy green vegetable referred to as Chinese broccoli.
1-17	Yu	Yu	Fish; also sounds like the word for abundance.
1-18	Yunbao	Yuanbao	The currency of old China; also refers to Chinese New Year's lucky dumplings.
1-18	Gao jee	Jiaozi	Boiled dumplings of northern China.
1-19	Nien go	Nian gao	Sweet, dense steamed glutinous rice cake that is eaten during Chinese New Year.

Chapter/ Page	Cantonese	Pinyin/ Mandarin	English
1-20	Gut jai	Ju zi	Tangerines.
1-20	Look yau	You zi	Pomelos, also known as Chinese grapefruit.
1-20	Ngow sing	Niu wei beng	Cow tail cookies, a regional Chinese New Year's treat.
1-20	Gok jai	—	Little half-moon-shaped cookies, a regional Chinese New Year's treat.
1-21	Fah Yuen	Hua Xian	Meaning "flower garden," a district in southern China's Guangdong Province.
1-25	Choon	Cong	Green onion; also sounds like the word for smart.
1-28	Yu san	Yu sheng	Raw fish salad eaten during Chinese New Year.
1-30	Tong yuen	Tang yuan	Rice flour dumplings that are eaten in either a sweet syrup or a savory soup.
1-30	Yuan Shiu	Yuan Xiao	Young servant girl and central character of the legend in which the tradition of eating *tang yuan* and the Lantern Festival was founded.
1-31	Lok dai hoi fa	Luo di kai hua	A saying to counteract bad luck when something breaks, meaning "Fall to the floor and open with flowers."
1-31	Sui sui ping on	Sui sui ping an	A saying to wish peace when something breaks that wishes peace for

Chapter/ Page	Cantonese	Pinyin/ Mandarin	English
			every year, because the word "broken" also sounds like the word "peace" in Mandarin.
1–32	Gum Lung	Jin Long	Golden dragon.
2–44	Ching Ming	Qing Ming	Clear Brightness Festival.
2–46	Hoon dynasty	Han dynasty	China's dynastic rule between 206 B.C.E and 200 C.E.
2–46	Heng san	Sao mu	Cantonese colloquial term for "walk mountain," referring to visiting the Chinese hillside cemetery to pay ancestral respects.
2–46	Feng shui	Feng shui	Meaning "wind-water," the Chinese philosophy and practice of harmonic placement and balance of things to complement life's energy.
2–51	Wan Bo	Wan Bo	Poet who won a literary competition in the folk tale where the practice of burning joss paper and otherworld money was founded to pay homage to spirits and heavenly deities.
2–51	Chang Lu	Chang Lu	Old man-spirit from the folk tale where the practice of burning joss paper and otherworld money was founded to pay homage to spirits and heavenly deities.

Chapter/ Page	Cantonese	Pinyin/ Mandarin	English
2-53	Guy Zi Tui	Jie Zi Tui	Jin state soldier and loyalist to the Duke of Wen who perished in a fire, which resulted in the duke's proclamation that no fires ever be set and only cold food be eaten on the eve of Qing Ming.
2-53	Gum kok	Jin state	Chinese state ruled by the Duke of Wen.
3-61	Wat Yuen	Qu Yuan	Beloved poet and statesman who perished by self-sacrifice on which the dragon boat festivities are based.
3-62	Lo ji	Lao zi	Lao-tzu, 6th-century B.C.E. Chinese philosopher and alleged author of the *Tao-te Ching (Daode Jing)*; considered the founder of Taoism.
3-63	Chong Kuai	Zhong Kui	Demon slayer whose picture can hang on the front door to avert evil.
3-65	Lei Sow	Li Sao	Qu Yuan's world-famous poem "On Encountering Sorrow," written over 2,000 years ago.
3-65	Mei lo River	Miluo River	Hunan Province river where poet and statesman Qu Yuan perished.
3-65	Yangtse River	Yangzi River	Mayor tributary in China that's considered the third longest river in the world. It travels 6,380 km

Chapter/ Page	Cantonese	Pinyin/ Mandarin	English
			(3,964.35 miles) from the Geladandong Glaciers of Tibet to the East China Sea.
3-66	Lung	Long	Dragon that is considered supreme and is the imperial symbol of the emperor.
3-66	Lei	Li	Water dragon that rules the rivers, seas, rain, and directions of the compass.
3-67	Gao	Jiao	Earth dragon whose energy is tapped when practicing the concepts of feng shui.
3-67	Mong	Mang	The dragon of commoners.
3-67	Joong	Zongzi	Packets of savory or sweet glutinous rice and other ingredients tied in bamboo leaves that are made during the Dragon Boat Festival in honor of the poet Qu Yuan.
5-89	Yuelon	Yulan	Taoist-Buddhist temple ceremony conducted during the Hungry Ghosts Festival.
5-89	Mu Lian	Mu Lian	Buddhist disciple who sought to free his mother from the land of hungry ghosts and to attain a place of peace during the seventh month.
5-90	—	Di Zang Pusa	God of inner nature and pure thought that resides within all human beings.
6-94	Chang-O or Shiang-O	Chang-E	The Moon Goddess.

Chapter/ Page	Cantonese	Pinyin/ Mandarin	English
6-96	Hau Yi	Hou Yi	The Divine Archer who is the Moon Goddess's husband.
6-98	Ng Gong	Wu Gang	The moon's woodcutter.
6-99	Yuet beng	Yue bing	Moon cakes that are served and given as gifts during the Mid-Autumn Festival.
6-99	Empress Dowager Ci Xi	Empress Dowager Ci Xi	The last empress of the Qing dynasty (1644–1911 C.E.).
6-99	Ching dynasty	Qing dynasty	Dynastic rule between 1644 and 1911 C.E.
6-101	Doe see	Douchi	Preserved salted black beans.
6-104	Liu Fu Tong	Liu Fu Tong	Han leader who developed the strategy of hiding messages inside moon cakes to unify the rebellion against the Mongol Yuan dynasty (1279–1368 C.E.).
6-104	Yuen dynasty	Yuan dynasty	Dynastic rule between 1279 and 1368 C.E., descendants of Genghis Khan.
7-108	Chung Yeung	Chong Yang	Day of the Double Sun, also known as Double Ninth Day.
7-111	Dang go	Deng gao	The practice of literally "ascending heights" during Double Ninth Day.
7-111	Woon King	Huan Jing	Han scholar who averted disaster by taking his family on an outing during

Chapter/ Page	Cantonese	Pinyin/ Mandarin	English
			Double Ninth Day. In another legend, he is the mortal who saved his villagers by evacuating them to higher ground and defeating the evil nemesis, the God of Plague, on Double Ninth Day.
7–112	—	Fei Zhangfang	A wise monk who taught Huan Jing the secrets of averting disaster to defeat the God of Plague.
8–125	Lai beng	Li bing	The practice of gifting bridal engagement cookies.
8–126	Beng	Bing	Generic name for an assortment of individual Chinese pastries that contain various fillings with a delicate outer crust.
8–127	Dim sum	Dian xin	A Chinese meal meaning "touch the heart" consisting of an assortment of small tidbits of food presented on roving carts in a restaurant or carried out.
8–130	Hong qua	Hong pao	Bride's red suit ensemble.
8–130	Cheongsam	Qipao	Chinese long fitted dress.
8–131	Mao dai	Yinger dai	Baby sling used as a carrier by Chinese mothers.
8–132	Yuk	Yu	Jade.

Chapter/ Page	Cantonese	Pinyin/ Mandarin	English
8-135	Full jook tong	Fuzhu tang	Dried bean curd soup.
8-135	Jum cha	Dao cha	The act of serving tea.
9-154	Chau yuet	Zuo yue zi	A new mother's sitting month.
9-154	Leurng	Hangqi	The "coolness" of a body's internal temperature.
9-154	Chi	Qi	Life energy.
9-154	Fung sup	Feng shi	Irreversible cold condition.
9-154	Yeet hay	Shang huo	The "hot breath" of a body's internal temperature.
9-155	Gai jow	Ji jiu	Chicken-wine soup served to new mothers.
9-155	Mun yurt	Man yue	"Full month" milestone of a new child's life.
9-159	Sern gueng	Suan jiang	Pickled ginger, which also connotes grandchildren and deep "ginger" roots that perpetuate a family's lineage.
9-159	Geen doi	Jian dui	Deep-fried sweet sesame balls served during dim sum.
9-160	Teem jow	Tian jiu	Fermented sweet rice wine pudding.
10-171	Dai san yat	Da shou	Big birthday.
10-172	Cheung so	Chang shou	"Seizing longevity," the act of celebrating a big birthday a year ahead of time.
10-173	Mah jong	Ma jiang	Chinese game played with four players at a table

Chapter/ Page	Cantonese	Pinyin/ Mandarin	English
			with decorative tiles. Tiles are drawn and discarded until one player wins with a hand of four combinations of three tiles each and a pair of matching tiles.
10-174	Shau toh	Shou tao	Small peach-shaped steamed bun with sweet bean paste that is served during big birthdays.
10-177	Yum bui	Gan bei	A Chinese toast, equivalent to saying "Bottoms up."
10-180	Shau Sing Gung	Shou Xing	The Star of Longevity or the God of Longevity.
10-181	Shau	Shou	Longevity.
10-182	Gum pi	Jin bian	Chinese characters, symbols, or images designed in 24-karat gold and encased in gold frames given as gifts for big birthdays, weddings, or business openings.
10-183	Ping gwa	Ping guo	Apple.
10-184	Luk	Lu	Synonym to prosperity and deer.
10-185	Gai don cha	Ji don cha	Brown tea-egg soup served during birthdays.
11-193	Toishan	Taishan	A southern district of Guangdong from where many of the Gold Mountain immigrants emigrated.
11-195	Fukien	Fujian	Southern China province.

Chapter/ Page	Cantonese	Pinyin/ Mandarin	English
11-194	Gum San	Jin Shan	"Gold Mountain"—term used by Chinese immigrants for California during the gold-mining era.
11-195	Look San	Zhonghua Zong Hui Guan	Meaning "Six Companies," which refers to the Consolidated Chinese Benevolent Association, the organizing entity for the many Chinese benevolent associations.
11-197	Pei	Bei	Blankets that the children of a deceased parent lay over a body in the casket as a gift of eternal warmth and comfort in the otherworld.
11-197	Shau pei	Shou pao	Longevity blanket for the deceased.
11-197	Fut pei	Fo bei	Spirituality blanket for the deceased.
11-198	Saam soong	San cai	Referring to the funeral food offering of "three dishes" that includes roast suckling pig's head, a whole white chicken, and *jai choy*—the "monk's" vegetarian dish.
11-199	Ha gow	Xia jiao	Shrimp dumpling of *dim sum* assortment.
11-199	Siu mai	Shao mai	Pork dumpling of *dim sum* assortment.
11-204	Kai ching	—	Yellow strips of devil money to distract evil spirits during a Chinese funeral.

Chapter/ Page	Cantonese	Pinyin/ Mandarin	English
11–206	Shau chaan	Shou can	Longevity dinner following a Chinese funeral.
12–219	Jung	Zhong	Word for "clock" that sounds like "funeral" in Cantonese, and "the end" or "termination" in Mandarin, and therefore prevents giving clocks as a gift.
12–225	Soong	Cai	A Chinese dish or entrée.
12–227	Canton	Guangzhou	Commonly referred to today as Guangzhou, the old port city in the Guangdong Province.
12–227	Peking	Beijing	China's capital, formerly known as Peking, the city of the old imperial dynasties.
12–228	Szechuan	Sichuan	Province of China known for its spicy cuisine; Chengdu is its capital city.
12–229	Mou Tai	Mao Tai	Popular toasting liquor of China.
12–229	Sancheng Chiew	—	Common rice wine used for chicken-wine soup for new mothers.
12–229	Goh Lerng	Kao Liang	Clear distilled spirit similar to vodka, from northern China.
12–229	Siu Heng	Shao Xing	Rice wine similar in color to sherry that originates from Zhejiang Province and is served warm like sake.
12–229	Ng Ka Py	Wu Jia Pi	Dark yellow spirit from Guangzhou that is similar to strong bourbon.

Chapter/ Page	Cantonese	Pinyin/ Mandarin	English
12-230	Mui Kwe Lu	Mei Gui Lu	Strong liquor distilled from grains that has an essence of rose petals.
12-230	Tsingtao	Xing dao	China's most popular beer.
12-230	Cha	Cha	Tea.
12-230	Sun Nong	Shen Nong	The Chinese emperor who is considered the father of agriculture and herbal medicine and is credited with discovering Chinese tea in 2737 B.C.E.
12-231	Nanking	Nanjing	Well-known capital city of the Jiangsu Province.
12-231	Ho Nam	Henan Province	Central China province; Zhengzhou is its capital city.
12-231	Lung Ching	Long Jing	Dragon Well green tea.
12-231	Hangjao	Hangzhou	Chinese city with West Lake and considered the summer home of the imperial court in Zhejiang Province.
12-231	Jiet Gong	Zhejiang Province	Eastern China province below Jiangsu and Shanghai. Hangzhou is its capital city.
12-232	Bik Lo Choon	Biluochun	Popular tea from Jiangsu Province that was a favorite of Chairman Mao.
12-232	Gong So	Jiangsu Province	Eastern coast Province on the East China Sea; Nanjing is its capital city.
12-232	Ping Sui Chue Cha	Pingshui Zhucha	Pearl Tea; also referred to as "Gunpowder" tea because it's rolled into round pellets.

Chapter/ Page	Cantonese	Pinyin/ Mandarin	English
12-232	Wong Shaan Mo Fung	Huangshan Maofeng	Famous green tea from Anhui Province.
12-232	—	Wu long	A semi-fermented group of teas commonly known as oolong, or "black dragon," in Chinese.
12-232	—	Tieguanyin	An oolong tea known as Iron Goddess of Mercy.
12-232	Hong cha	Hong cha	Meaning "red tea," the Chinese reference for black tea.
12-232	—	Keemun	High grade of black tea from Anhui Province.
12-233	Po nei, Po nay, or Bo lay	Pu'er or Pu-erh	Exotic, earthy dark tea that is fully fermented and classified as a black Chinese tea.
12-233	Soong dynasty	Song dynasty	Dynastic rule between 960 and 1279 C.E.
12-233	Yee Hing	Yixing	City near Shanghai in Zhejiang Province that is well known for its clay teaware.

bibliography

Ball, J. Dyer. *Things Chinese: Or Notes Connected with China*. New York: Oxford University Press, 1982.

Barker, Pat. *Dragon Boats: A Celebration*. New York: Weatherhill, 1996.

Blonder, Ellen. *Dim Sum: The Art of Chinese Tea Lunch*. New York: Clarkson Potter, 2002.

Blonder, Ellen, and Annabel Low. *Every Grain of Rice: A Taste of Our Chinese Childhood in America*. New York: Clarkson Potter, 1998.

Bredon, Juliet, and Igor Mitrophanow. *The Moon Year: A Record of Chinese Customs and Festivals*. New York: Paragon Book Reprint Corp., 1966.

Burkhardt, V. R. *Chinese Creeds and Customs*. Hong Kong: South China Morning Post, 1982.

Chang, Iris. *The Chinese in America*. New York: Viking Penguin, 2003.

Chen, Huoping, Tan Huay Peng, and Leong Kum Chuen (illus.). *Fun with Chinese Festivals*. Union City, Calif.: Heian International, 1991.

Chow, Kit, and Ione Kramer. *All the Tea in China*. San Francisco: China Books and Periodicals, 1990.

Costa, Shu Shu. *Wild Geese and Tea: An Asian-American Wedding Planner*. New York: Riverhead Books, 1997.

Dotz, Warren, Jack Mingo, and George Moyer. *Firecrackers: The Art and History*. Berkeley, Calif.: Ten Speed Press, 2000.

Eberhard, Wolfram. *A Dictionary of Chinese Symbols*. New York: Routledge, 1986.

Feng, Doreen Yen Hung. *The Joy of Chinese Cooking*. New York: Grosset & Dunlap, 1992.

Fong-Torres, Ben. *The Rice Room: Growing up Chinese American from Number Two Son to Rock 'n' Roll*. New York: Hyperion, 1994.

Foster, Dean. *The Global Etiquette Guide to Asia*. New York: John Wiley & Sons, 2000.

Hu, William C. *The Chinese Mid-Autumn Festival: Foods and Folklore*. Ann Arbor, Mich.: Ars Ceramica, 1990.

———. *Chinese New Year: Fact and Folklore*. Ann Arbor, Mich.: Ars Ceramica, 1991.

Kan, Johnny, with Charles L. Leong. *Eight Immortal Flavors*. Berkeley, Calif.: Howell-North Books, 1963.

Keltner Wong, Kim. *The Dim Sum of All Things*. New York: HarperCollins Publishers, 2004.

Ki, Goh Pei. *Origins of Chinese Festivals*. Singapore: Asiapac Books, 1997.

Kingston, Maxine Hong. *China Men*. New York: Alfred A. Knopf, 1997.

———. *The Woman Warrior: Memoirs of a Girlhood Among Ghosts*. New York: Alfred A. Knopf, 1976.

Lip, Evelyn. *Chinese Practices and Beliefs*. Torrance, Calif.: Heian International, 2000.

————. *Choosing Auspicious Chinese Names*. Torrance, Calif.: Heian International, 1997.

Lo, Eileen Yin-Fei. *The Chinese Banquet Cookbook*. New York: Crown, 1985.

McCunn, Ruthanne Lum. *The Moon Pearl*. Boston: Beacon Press, 2000.

Min, Anchee. *Wild Ginger*. New York: Houghton Mifflin, 2002.

Qu, Yuan. *The Li Sao: An Elegy on Encountering Sorrows*. Translated by Boon Keng Lim. Shanghai: The Commercial Press, 1929.

Robinson, Fay. *Chinese New Year: A Time for Parades, Family and Friends*. Berkeley Heights, N.J.: Enslow Publishers, 2001.

Shan, Lin. *Name Your Baby in Chinese*. Union City, Calif.: Heian International, 1988.

Sharp, Damian. *Simple Chinese Astrology*. Berkeley, Calif.: Conari Press, 2000.

Simonds, Nina. *Chinese Seasons: A Celebration of Classic and Innovative Chinese Dishes*. Boston: Houghton Mifflin, 1986.

Stepanchuk, Carol. *Red Eggs and Dragon Boats: Celebrating Chinese Festivals*. Berkeley, Calif.: Pacific View Press, 1994.

Stepanchuk, Carol, and Charles Wong. *Mooncakes and Hungry Ghosts: Festivals of China*. San Francisco, Calif.: China Books and Periodicals, 1991.

Stepanchuk, Carol, and Leland Wong (illus.). *Exploring Chinatown: A Children's Guide to Chinese Culture*. Berkeley, Calif.: Pacific View Press, 2002.

Sung, Vivien. *Five Fold Happiness*. San Francisco: Chronicle Books, 2002.

Tam, Vivienne. *China Chic*. New York: HarperCollins Publishers, 2000.

Tan, Amy. *The Joy Luck Club*. New York: G. P. Putnam's Sons, 1989.

————. *The Kitchen God's Wife*. New York: G. P. Putnam's Sons, 1991.

Ward, Barbara, and Joan Law. *Chinese Festivals in Hong Kong*. South China Morning Post Ltd Publications Division, 1982.

Warner, John. *Chinese Papercuts*. Hong Kong: John Warner Publications, 1978.

Xing, Qi, ed., and Yang Guanghua (illus.). *Folk Customs at Traditional Chinese Festivities*. Translated by Ren Jiazhen. Beijing: Foreign Languages Press, 1988.

Yan, Martin. *Martin Yan's Chinatown Cooking*. New York: HarperCollins, 2002.

Yang, Jeff, Dina Gan, Terry Hong, and the staff of *A. Magazine*. *Eastern Standard Time: A Guide to Asian Influence on American Culture from Astro Boy to Zen Buddhism*. Boston: Houghton Mifflin, 1997.

Zhenyi, Li. *100 Ancient Chinese Customs*. Translated by Yao Hong. Hong Kong: The Commercial Press, 1996.

acknowledgments

After attaching my final chapter of *Good Luck Life* to an e-mail to HarperCollins, I clicked the "send" button and held my breath in anticipation of the words "File's done." It took a community of family, friends, business associates, and social acquaintances to reach that moment, and I'm indebted to them for their generosity of time, information, and advice.

I'm grateful to Edwin Tan, editor extraordinaire, of HarperCollins. Highly intuitive and partly clairvoyant, Edwin envisioned *Good Luck Life* in print before I did. With his unwavering enthusiasm and culturally evolved eye, he gently probed and prodded. Edwin made my publication process effortless. And as a bonus, I received a Straits-style recipe for a new twist on *joong*. I'm also indebted to Jessica Chin, production editor at HarperCollins, for her wonderful eagle eye.

It was my great fortune to be blessed with an inspired and generous foreword by master chef Martin Yan. His tireless devotion and commitment to Chinese food and culture sets an example for us all.

As a first-time author, I was shown the ropes by my literary agent, Diane Gedymin. She laid the groundwork and spoke the truth. When Diane transitioned back to publishing, she lovingly surrendered me into the prodigious hands of Michael Carlisle.

Paola Gianturco has continued to mentor me beyond my days at Saatchi & Saatchi Corporate Communications Group. I never thought that I would follow her into the world of publishing. But when *Good Luck Life* was just a wisp of an idea, it was Paola who told me that I could do it and gave me the map that showed me how.

Through Chinese experts Aubrey Kuan of the Monterey Institute of International Studies, Lillian Hwang Peiper of Crane House—The Asia Institute, Inc., and Sally Yu Leung, San Francisco Asian Art Commissioner, I learned about the details and nuances of the Chinese ways with regional overtones among east and west, north and south, Taiwanese and overseas Chinese. They patiently reviewed, critiqued, and edited my chapters and provided me with a depth of knowledge that could not be found in any book.

The support and diligence of my writing group, the Random Writers, consisting of Sara Bir, Kirk Citron, Leslie Van Dyke, and Chun Yu, kept me focused and steady. Thanks for the DBs, SMs, WCs, and LOLs in colored ink. Sara Bir also contributed her culinary talents by meticulously testing and refining the old village recipes in her own kitchen.

Wylie Wong, Jeannie Young, and Jane Huie Lang bestowed their magic. They granted wishes at pivotal times when I couldn't manifest them on my own. My personal editor, Jennifer Tansey Greenwood, was my guardian angel. During the writing process, she kept me focused with the buoyancy of a life jacket.

I was fortunate to meet the San Francisco Bowlers, an informal sorority of Chinese American mothers who have been bowling together since 1965 and who happened to rear two generations of Chinese Americans between strikes. My thanks to Ella Chan, Fannie Chinn, Jessie Eng, Irene Gee, Betty Hum, Bernice Louie, Helen Lum, and Hattie Wong, with a special thanks to Betty Leong Louie, who provided her personal collection of books and clippings. Other wise uncles and aunties who contributed their knowledge were Cain and Yvonne Chan, Lola Chan, Helen Lang, David Lowe, Frank Mah, Suzanne Pan, Ruby Pao, and May Young. The Chinese aunties who provided their celebratory family recipes were Linda Chan, Peggy Chu, Mary Gong, Susie Huie, Lynn Lowe, and Ruby Young.

It was Linda Cheu and Hans Wu of the California Dragon Boat Association who taught me about the sport and competition of dragon boat racing. Ida and Willard Lee, former owners of the Great Eastern Bakery, provided insights on the variations of moon cakes and bridal cookies. Doreen Chin explained Shanghai cuisine and specialties. David Wong of Imperial Tea Court provided his tutelage on Chinese tea and assisted me with the Chinese languages. Stephan Chan of San Francisco's Empress of China Restaurant graciously offered the traditional bride's sweet lotus seed soup recipe that appears in the wedding chapter.

For information and instructions on the Chinese funeral, I'm indebted to Bill Steiner and Samson Law of San Francisco's Green Street Mortuary and Clifford Yee of Cypress Lawn Memorial Park. Robert

Wong of Chi Sin Temple explained the Ghost Festival, *Yulan,* temple ceremony, and Wayman Dea of All Seasons Flowers provided a primer on Chinese flower arrangements.

I am honored that John Way, an elegant, highly schooled master calligrapher, felt that *Good Luck Life* was worthy of his calligraphy for the chapter headings. I owe a debt of gratitude to Bob Pimm for his legal eye and Maurice Chuck for his advice and expertise on the glossary.

Many thanks to Elsie Wong, branch manager, and the reference desk librarians at the San Francisco Public Library's Chinatown branch, who helped identify resources to answer my mountain of questions. They verified facts, helped with Chinese words and translations, and cheered me on. The staff and resources at the Chinese Historical Society and Chinese Culture Center were also invaluable.

My teachers who provided their insights on the writing craft and publishing were John Darling, Constance Hale, Leslie Keenan, Cathy Luchetti, Linda Watanabe McFerrin, Stephanie Moore, and Peggy Vincent. I cannot say enough good things about the Left Coast Writers Literary Salon and the community of writers and teachers at Marin County's Book Passage. My single word of advice for any budding writer in the Bay Area is *go.*

To the Asian American writers who encouraged me to pursue *Good Luck Life* with their personal words of wisdom—Ellen Blonder, Tess Uriza Holthe, Maxine Hong Kingston, Ruthanne Lum McCunn, and Gail Tsukiyama—thank you for your inspiration.

Valerie Wong of Wong and the Design Office was always ready and willing to give a designer's graphical perspective as *Good Luck Life* developed. For this, I also thank Greg Chew, Carin Christensen Fleming, Scott Giusti, Perry Lucina, Dave Sanchez, and Han Vu. I'm grateful to Mitzi Ngim for my Website at www.goodlucklife.com.

Seismicom is my professional family. I consider the encouragement, freedom, and security that Seismicom showed me exceptional. They allowed me to explore something new and minimized my risk by offering me a place to which I could return. Thanks to partners Bill Carmody, David Flaherty, and Kathy Mitchell, and the Seismicom staff, who honored me with my very own Chinese tea party.

Joyce Yokomizo provided humor, friendship, and my author's photo—with Keiko Ito's digital assistance. I'm indebted to Kenny Watanabe and Meredith Linamen for high-resolution photography scans. Ainslee Faust gave me her encouragement and support, as did the Last Thursday Writers Group that included Daniel Chouinard, Philip Cohen, Rochelle Frey, Kia Jewels, Charli Ornett, Emily Mitchell, Alex Momtchiloff, Rafael Olivas, Mark Richardson, Vanessa Richardson, and Bill Schroeder.

My aunt Sherli took it upon herself to pick me up and dust me off. Thank you for the strong black coffee and chorizo scrambles. Thanks also to *Rosebud*, the blazing energy that hangs on my wall, by my friend, painter Susan Nettelbeck.

It was my siblings, Becky, Tommy, and Lori, who indulged my varied mood swings. My sister-in-law Sherry showed me the importance of reading and writing Chinese. Thanks to my two brothers-in-law—Terry, who could have written this book, and Brian, who'll learn from it—for loving my sisters. To my four nephews who are the family jewels, Brandon, Ryan, Derek, and Darin, thank you for making me smile.

During the months that I devoted my full energies to *Good Luck Life*, my family had a seventieth birthday, a new baby boy, a wedding, and a funeral. At that wedding, as I looked out across the sea of banquet tables at my extended family—the Chins, Chus, Laus, Lowes, Gongs, Kwocks, and Youngs—I had to pinch myself because I felt so lucky. *Good Luck Life* is because of them.

index

agnate ancestors, 49
Alcan Dragon Boat Festival, 75
alcoholic beverages, 229–30
  see also liquors; wines; *specific drinks*
altars, 88
  home, 14, 16, 82, 102, 124
  for weddings, 137
American Dragon Boat Association, 75
ancestors, 14, 46, 108, 163
  agnate, 49
  in Chinese New Year, 16, 18, 29
  in Chong Yang, 108, 110
  in Clear Brightness Festival, 45–50, 52,
    54
  in Mid-Autumn Festival, 102
  and naming of babies, 163
  presentation of infant to, 158
  in wedding ceremony, 138
apples, 47, 102, 136, 184
armbands, 200

art, decorative, 183–84
Asian Dragon Boat Federation, 76
astrological animal years, xvi, 33–40,
    123–24, 162, 165, 171
attire:
  for mourning, 200
  for New Year, 13–14
  for Red Egg and Ginger Party, 168
  for weddings, 129–31
Auntie Lao, 19, 30, 49, 52, 64, 114, 125,
    132, 233
  folk wisdom of, 52, 104, 123, 141,
    152, 157, 160, 165, 166, 167, 183,
    197, 200, 205, 224
Auntie Lynn's engagement sponge cakes,
    128–29
Auntie Peggy's *gok jai* cookies,
    22–24
Auntie Ruby's black vinegar pigs' feet,
    156–57, 159

bamboo, bamboo leaves, 67, 68, 110
banners, 203, 204
benevolent associations, 193–95, 204
beverages:
    alcoholic, 229–30
    tea, 230–34
    see also specific beverages
big birthday (*dai san yat*), 169–88
big birthday banquet, 173–88
    menu for, 174–75
    planning guide for, 173, 186–88
Big Dipper, 105
Biluochun, 232
bird's nest soup, 16, 141, 174, 225
birthdays, 90, 162
    big, 169–88
    cake for, 179
    in Chinese New Year, 26–28
    favors for, 180
    first, 166–67
    infants' one-month, 158–60
    invitations for, 175
    sixtieth, 171
black, symbolism of, 13, 31, 175, 200,
    210, 219
black bean sauce, sautéed snails with,
    101
black tea, 232
black veils, 200
black vinegar pigs' feet, Auntie Ruby's,
    156–57
blanket(s) (*pei*), 197
    of spirituality (*fut pei*), 197
blue, 210, 219
boar, year of, 39
Bodhidharma, 231
Boublil, Alain, 120
bowing, 49, 102, 110, 138, 158, 202,
    209
Bredon, Juliet, 50
bride's cookie day (*lai beng*), 125–26
bridesmaids, 135
brown tea-egg soup, Mom's, 185–86
Buddha, 89, 90, 98
Buddhism, Buddhists, 89, 90, 91, 98,
    197, 201, 202, 231
burial, 204–5
butterflies, 132, 134

cakes, Auntie Lynn's engagement sponge,
    128–29
calendar, Chinese, xv–xvi, 33–34, 193
California Dragon Boat Association, 76
camellia, 11
Camellia sinensis, 231
candies, 19
candles, 102, 110
Cantonese cuisine, 227
care packages, 206–7
cassia tree, 97, 98, 102
"catching rain," 55
cats, 33
cemetery rituals, 46–51, 55–57, 109–10,
    208–10
centipedes, 63
Chan, Linda, 21
Chang-E (Moon Goddess), 94, 95,
    96–97, 103
Chang Lu spirit, 51
charitable contributions, 140, 178–79
charms, protection, 54, 63–64, 112
checklists:
    for Clear Brightness Festival, 55–56
    for first cemetery visit, 208–9
cheongsam, 129, 130–31, 144
chestnuts, 26
chicken, 47, 100, 110, 198, 206,
    208
    -wine soup (*gai jow*), 155–56, 159
children, 14, 15, 167
China Men (Kingston), 158–59
Chinese New Year, xvi, 3–41
    "birthdays" in, 26–28
    dates of, 40
    do's and don'ts for, 30–31
    eve of, 14–18
    fifteen days of, 26–28
    flowers for, 11–12
    funerals and, 193
    greetings for, 15
    parade, 31–32
    planning guide for, 39–41
Chinese New Year's Eve, 14–18
    dinner, 16
    menu for, 16–18
Chong Yang (Day of the Double Sun),
    107–15

chopsticks, 18, 139, 140, 180, 198, 222–24
   Chinese vs. Japanese, 222–23
   etiquette for, 223–24
chrysanthemums, 12, 110
Chu, Peggy, 22
Chu State, 64
citron, 11
Ci Xi, Empress Dowager of China, 99, 192
Clear Brightness Festival (Qing Ming), 43–57, 109
   checklist for, 55–56
   four steps of, 46–51
   planning guide for, 57
clocks (*jung*), 219
coconuts, 25, 102
coins, 184, 192
colors:
   imperial, 67
   see also specific colors
combs, 197
Confucianism, 46, 123, 172, 210
Consolidated Chinese Benevolent Association (*Look San*; Six Companies), 195
constellations, 81
cookies, 19, 20, 99
   Auntie Peggy's *gok jai*, 22–24
   day, bride's, 125–26
   Mrs. Chan's cow tail, 21–22
corsages, boutonnieres, 139, 176
cow, 80
Cowherd and the Weaving Maiden, 79–82
cow tail cookies, Mrs. Chan's, 21–22
cow tails (*ngow sing*), 20, 21–22
cranes, 180, 184
cuisine, Chinese, 226–29
   Cantonese, 227
   Hunan, 228–29
   Mandarin, 227
   Shanghai, 228
   Sichuan (Szechuan), 228
   for weddings, 140–41
cursing, 31
cypress, 11

dead:
   household goods for, 198–99
   journey of, 192, 196–97, 198, 207–8
   judgment of, 207–8
   rituals for, 43–57, 85–91, 109–10, 189–214
death, colors symbolic of, 12, 13, 31, 130, 192, 196, 197, 200, 210, 219
   see also funerals
debts, 8
deer, 180, 184
demons, *see* spirit(s), evil
deng gao (ascending heights), 111
dessert, 102
devil money (*kai ching*), 204
di (key to inner nature), 91
dimes, 19
dim sum, 127, 135, 198, 202, 218, 233
Dim Sum of All Things, The (Keltner), 199
dinner:
   filling glasses at, 222
   in the home, 220–21
   place settings for, 222
   in a restaurant, 221–22
   seating at, 221–22
Divine Archer, Ruler of the Solar System, 96, 97
Di Zang Pusa, 90–91
dog, year of, 38–39
donations, *see* charitable contributions
Dong Fang, 30
Door Gods, 10
Double Fifth, 61–62
Double Happiness, 126, 139
Double Ninth Day, 50, 107–115
Double Seventh Day (Seven-Seven), 77–83
dowry, 126, 129
dragon, year of, 36
Dragon Boat associations, 75–76
Dragon Boat Canada, 76
Dragon Boat Festival, 59–76
dragon boat races, 72–76
dragons, 32, 122, 130, 139
   golden, 32
   Jiao (earth), 66, 67
   Li (water), 66–67
   Long (imperial), 66

dragons (*continued*)
  Mang (common), 66, 67
  river, 62, 65, 67
"Drink cup" (toast), 177
drums, 33, 61, 65, 74, 105, 179, 201, 203
duck, 100, 141, 170, 208
dumplings, 18–19, 30

earth, 124
Earth God, 33, 52, 53
Eastern Regional Dragon Boat Federation, 75
egg-brown tea soup, Mom's, 185–86
"egg roll cookies," 20
egg yolks, 99, 104
eight dispositions, 124
Eight-Fifteen, 95, 96, 105
Eight Immortals, 31, 182, 184
elders, 14, 68, 138, 158, 161, 164, 172, 178, 181, 218, 220, 222
elixir of immortality, 96–97
embalming, 196
engagement sponge cakes, Auntie Lynn's, 128–29
entrance packets, 200–201
etiquette, 215–34
  for chopsticks, 223–24
  in gift-giving, 219–20
  at New Year, 15
  in restaurants, 221–22, 224–26
  at table, 217–18
European Dragon Boat Federation, 76
Everybody's Birthday, 172, 173
exit packets, 200–201

family altars, *see* altars, home
fat choy (sea moss), 17, 25
Fei Zhangfang, 112
feng shui, 46, 67
ferry, 196
festivals:
  Clear Brightness, 44–57
  Dragon Boat, 60–76
  of Hungry Ghosts, 86–91

of Mid-Autumn "Moon" (Mid-Autumn Festival), 94–106
fifteen days of Chinese New Year, 26–28
filial piety, 123, 172, 210
financial responsibilities:
  for big birthday banquet, 174
  for weddings, 139
finger tapping, 226
fire, 53–54
Fire, God of, 30
firecrackers (*pau jeun*), 13, 14, 32, 51, 73, 137, 138
first cemetery visit, 208–9
first full moon, 29
fish, 17, 25, 141, 170, 184, 208
five blessings, 183
five elements, 124, 162, 171
Five Gods of Plague (five evil gods), 62, 63–64
five points of direction, 67, 69
five poisonous creatures, 63
five treasures, 126
flashlight, 207
flavored tea, 233
floating lanterns, 88
Flower Fair, Chinese New Year, 12
flowers:
  for big birthday banquet, 176
  in Chinese New Year, 11–12
  in Chong Yang, 110
  in Clear Brightness Festival, 47, 49, 53, 56
  in Double Ninth Day, 113
  in Dragon Boat Festival, 64
  in funeral services, 195–96, 205
  as hostess gifts, 219
  for months, 176
  for seasons, 177
food:
  for big birthday banquet, 174–75
  in bride's cookie day, 126
  in Chinese New Year, 16–20
  choice in restaurant, 224–25
  in Chong Yang, 110
  in Clear Brightness Festival, 47–48, 50, 53
  in Double-Seventh Day (Seven-Seven), 82

in Dragon Boat Festival, 67–72
for first cemetery visit, 208–9
as funeral offerings, 198, 199, 208
as hostess gifts, 218
in Hungry Ghosts Festival, 88
for longevity dinner, 206
in Mid-Autumn Festival, 100–105
in *mun yurt* dinner, 159–60
for Red Eggs and Ginger parties, 164
in wedding banquet, 140–41
food decoder, 25–26
fortune telling, 82–83, 114, 124, 143, 166–67
fortune tray, 166–67
Foshan dragon masters, 32
"four gentlemen" of the garden, 110
frog, 97
fu (fook), 9, 10, 210
full jook tong, 135
full month (*mun yurt*), 158–59
funeral homes, 193, 196, 197, 201
funerals, 12, 14, 175, 189–213
    Christian elements in, 197, 202–3
    planning guide for, 211–13
    processions in, 203–4
fung sup, 154

gates of justice, 198, 207
gender, 153, 161, 162, 163, 201
generation names, 163
Genghis Khan, 104
Gewürztraminer, 177
ghosts, 86–91, 123, 201
    see also spirit(s)
gifts, gift-giving, 218–20
    for betrothal, 125
    for birthdays, 181–84
    for business acquaintances, 218–19
    for hostess, 218–20
    for Red Egg and Ginger parties, 164–65
    returning of, 125, 127
ginger, 154, 157, 159, 160, 162
given names, 161–62
glasses, refilling of, 222

glutinous rice cake (*nian gao*), 19–20
gok jai cookies, 20, 22–24
    Auntie Peggy's, 22–24
gold, 130, 197, 210, 219
    images, 181, 182
golden dragon, 32
golden staff, 90
Good Fortune Tray, 129
good luck, *see* luck, good
good wishes, 9
gossiping, 31
gourds, 102
grave site, 47
Great Healer, 54
green, 200, 219
Green Dragon Sword, 112, 113
green onions, 25
greens, leafy, 17
green tea, 231–32
greetings, 15, 217
Gum Lung (golden dragon), 32
gum pi (gold images), 181, 182

haggling, 132
handkerchiefs, 219
Han dynasty, 46, 104, 111
harvest moon, 99, 104
"Have you eaten yet?" 217
Heaven, Queen of, 113
Heavenly Mother, 81
heng san (walk mountain), 46, 208–9
    ritualistic kit, 48
herb of immortality, 180
hiking, 111
Hong Kong Dragon Boat Association, 76
hong qua (red suit), 129, 130, 143
hope chest, 129
horse, 102–3
    year of, 37
hostess gifts, 218–20
Hou Yi, 96
Huangshan Maofeng, 232
Huan Jing, 111–13
hubris, 96
Huie, Susie, 69

Hunan cuisine, 228–29
Hungry Ghosts Festival, 85–91

imperial colors, 67
incense, 48, 49, 55, 73, 88, 89, 102, 110,
    113, 158, 198, 201, 202, 204, 209
infants, infancy, 29, 152–68
    celebration of firsts for, 158
    mun yurt and, 158
    naming of, 160–63
    Red Egg and Ginger Party for, 151–65
ingots, 18, 48–49, 184
International Dragon Boat Federation, 73,
    75

jade (*yuk*), 31, 132, 165
Jade Emperor, 6, 7, 30, 96, 98
Jade Rabbit, 97, 99
jai choy, 16, 198, 202, 206
jasmine tea, 233
jewelry, 131–32, 138
    for babies, 165
    as birthday gift, 182
    in wedding ceremony, 131–32
Jiao, 66, 67
Jie Zi Tui, 53–54
joong (zongzi), 67–72
    Mrs. Huie's savory, 69–72
joss paper, 7, 48–49, 51–52, 55–56, 88,
    89, 110, 113, 201, 202, 209

Kao Liang, 229
Keemun, 232
Keltner, Kim Wong, 199
Kingston, Maxine Hong, 86, 159
kiss, 140
Kitchen God, 6–7
knives, 31, 219
kumquats, 11, 25

lai beng, 125
Lantern Festival, 26, 29, 30
lanterns, 102, 105
Lao-tzu, 62

Lapsang Souchong, 232
lazy Susan, 225
leaves, 64
leftovers, 18
leurng (cool), 154
Li, 66–67
lily buds, 17
lion dancers, 31, 32–33, 140, 173,
    179
liquors, 47, 110, 177
Liu Fu Tong, 104–5
lizard, 63
Long, 66
longan fruit, 19, 25
longevity:
    blanket (*shau pei*), 197
    bun (*shau toh*), 174, 175
    dinner (*shau chaan*), 206
    robe (long-life robe), 175–76, 192,
        196
    symbols, 184
    vigil, 14
Longevity, God of (Shou Xing), 180,
    181, 182, 184
Long Jing, 231
long-life noodles, 174, 175
lotus, lotus flowers, 12, 88
lotus seed(s), 17, 25
    soup, sweet, 142
"love letters," 20
Lowe, Lynn, 128
luck, bad:
    Double Fifth as, 62
    at New Year, 31
    in number four, 20, 115, 218
luck, good:
    charms for babies as, 165
    as Double Ninth Day, 113
    fu (fook) as, 9, 10, 210
    at New Year, 10, 15, 16–19, 31
    numbers and, 114–15
    red eggs as, 159
lucky dumplings, 18–19
lucky foods, 25–26
lucky money, 14, 20, 33, 135, 166, 180,
    181–82
"lucky papers," 9
lunar calendar, xv

Lychee Black, 233
lychee nuts, 25, 26

magpies, 79, 81
maiden's ritual, 82
Maltby, Richard, Jr., 120
Mandarin cuisine, 227
Mang, 66, 67
mao dai, 131
mao tai, 177, 229
marching bands, 31, 204
marriages, 15, 29, 82
    arranged, 122–23
    see also weddings
martial arts, 33, 105, 179, 231
matches, lighters, 56
matchmaker, 122, 124
McCunn, Ruthanne Lum, 136
meatballs, 25
melon seeds, 19, 25
memory slate, 208
metal, 124
Mid-Autumn Festival, 93–106
Milky Way, 81
Miluo River, 65
Min, Achee, 162
Ming dynasty, 105
mirror, 64
Miss Chinatown USA pageant, 31–32
Mitrophanow, Igor, 50
Mom's brown tea-egg soup, 185–86
money:
    in bride's cookie day, 126–27
    devil, 204
    in dowry, 129
    lucky, 14, 20, 33, 135, 166
    see also otherworld money
Mongols, 104, 105
monkey, year of, 37–38
Monkey King, 31
moon, 94–106
moon cakes (*yuet beng*), 99, 102, 104–5
Moon Festival, *see* Mid-Autumn Festival
Moon Goddess (Chang-E), 94, 95,
    96–97, 103
Moon Hare, 97
Moon Minister of Marriage, 29, 104

Moon Palace, 97
Moon Pearl, The (McCunn), 136
Moon Year, The (Bredon and Mitrophanow),
    50
mothers, new, 154, 157
mourning time, 210
Mrs. Chan's cow tail cookies, 21–22
Mrs. Huie's savory *joong*, 69–72
mugwort, 64
Mui Kew Lu, 230
Mu Lian, 89–90
mun yurt (full month), 158–60
mushrooms, black, 17, 25

names, Chinese, 160–64
narcissus, 11
needles, needlework, 81–83
new mothers, 154, 157
New Year, *see* Chinese New Year
New York City International Dragon Boat
    Race Festival, 75
Ng Ka Py, 229
Nian, 13
nian gao, 19–20, 26
Nine-Emperor God, 113
noodles, 17, 25, 170, 174, 175, 192
Northern California International
    Dragon Boat Championships, 75
numbers:
    eight, 17, 18, 115, 181, 218
    four, 20, 115, 218
    nine, 114, 115, 135, 140–41, 174,
        181
    odd and even, 20, 48, 63, 115, 206,
        218
    repetitive, 114
    significance of, 114–15

Old Man of the South Pole, *see* Longevity,
    God of (Shou Xing)
omens, 82
"On Encountering Sorrow" ("Li Sao")
    (Qu Yuan), 65
onions, 64
oolong tea, 232
oranges, 17, 19, 20, 26, 47, 110

orchids, 11, 110
otherworld money, 48, 51–52, 55, 88, 89, 110, 201, 202, 209
"otherworld spirit" shops, 48
ox, year of, 34
oysters, 17, 26

Pacific Dragon Boat Association, 75
paper replicas, 198–99, 202
parade, Chinese New Year, 31–32
pastries, 125–26, 127
patron saints, 81–82
paying respect:
  in Clear Brightness Festival, 49–50
  to Earth God, 53
peaches, 11, 26, 102, 175, 182, 184
peach of immortality, 180
peanuts, 19, 26, 102
  yummy, 103
pearls, 132
peony, 11, 184
phoenix, 122, 130, 139
photographs, 204
pig(s):
  cookies, 99
  feet, Auntie Ruby's black vinegar, 156–57
  roast suckling, 126, 141, 198, 206, 208
pine, 11, 184
pineapple, 9–10, 26
pink, 31, 130, 196, 219
Plague, God of, 111–13
planning guides:
  for big birthday banquet, 173, 186–88
  for Chinese New Year, 39–41
  for funerals, 211–13
  for Mid-Autumn Festival, 106
  for Red Egg and Ginger Party, 167–68
  for weddings, 143–50
plants, 219
plum blossom, 12, 110
pockets, 196
poetry, 111
pomegranates, 26, 102
pomelos (Chinese grapefruit, *look yau*), 20, 100

pork buns, 47
pork roast, 47, 100, 110, 208
Portland-Kaohsiung Sister City Association Dragon Boat Races, 75
pregnancy, 152, 205
prosperity symbols, 184
Pu-erh tea, 233
puns, 183–84
purple, 175

qi (life energy), 154
Qing dynasty, 99
Qing Ming, *see* Clear Brightness Festival
Qingti, 89–90
Qin Shu Bao, 10
Qin state, 64
quarrels, 8
quinces, 11
Qu Yuan, 53, 61, 64–65, 67, 73, 74

rabbit(s), 98
  year of, 35
rain prayers, 55
rat, year of, 34
realgar, 63
rebuses, 183–84
red, symbolism of, 10–11, 14, 15–16, 130, 196, 197, 210, 219
red candles, 14, 48, 49, 55, 201, 202, 209
red dates, 19, 26
redeemer, 90
Red Egg and Ginger parties, 151–66
  "announcement" packages for, 164
  planning guide for, 167–68
red eggs, 158, 159
red envelopes (*lai see*), 15, 20, 33, 126, 127, 129, 135, 138, 166, 179, 180, 181, 182, 201, 205
red silk table cloth, 139, 168, 178
red tea (*hong cha*), 232
reincarnation, 90, 207–8
remarriage, 198
restaurants, 224–25
ribbon dances, 105

rice, 17, 25, 47, 65, 67, 100, 192
    bowl, 180, 225
Riesling, 177
river dragon, 62, 65, 67
river of death, 192
rooster, year of, 38

saam soong (three dishes), 198
Sakyamuni, 89
Sancheng Chiew, 229
San Francisco:
    Chinese New Year Flower Fair in, 12
    Chinese New Year's Parade in, 32
sautéed snails with black bean sauce, 101
scissors, shears, 31, 56, 219
scorpions, 63
sedan chairs, 121, 136, 137
Seven-Fifteen, 87–88, 89
Seven-Seven (Double Seventh Day), 77–83
Seven Sisters, 82
Seven-Thirty, 90
Shanghai cuisine, 228
Shaolin, 231
Shao Xing, 229
shark's fin soup, 16, 141, 147, 225
sheep, year of, 37
Shen Nong, Emperor of China, 230
shoes, removal of, 220
shou (longevity), 181, 184
siblings, 161, 163
Sichuan (Szechuan) cuisine, 228
signs, astrological, 34–39
silver, 130
Silver Needle, 233
"sin eaters," 202
sitting month, 154
snacks, 19
snails, 100
    sautéed, with black bean sauce, 101
snake(s), 63
    year of, 36
solar calendar, xv–xvi
soups, 16, 224
    chicken-wine, 155–56
    Mom's brown tea-egg, 185–86
    sweet lotus seed, 142
soybean plants, 102

speeches, 178
spider boxes, 83
spirit(s), 51–52, 109, 198–99
    evil, 10, 11, 13, 32, 51, 54, 63, 64, 90,
        136, 137, 204
    money, *see* otherworld money
    world offerings, 48–49, 158, 208–9
    see also ghosts
sponge cakes, 47
    Auntie Lynn's engagement, 128–29
spring couplets, 9
Spring Festival, *see* Chinese New Year
"Star of Longevity," 180
summer solstice, 61, 62
"Sun and Moon" from *Miss Saigon*
    (Boublil and Maltby), 120
Sun God, 29, 80, 81
surnames, 160–61, 163
sweep away, 8
sweet lotus seed soup, 142
sweet rice flour dumplings (*tang yuan*), 30

tablecloth, red silk, 139, 168, 178
table etiquette, 215–26
Taizong, Emperor of China, 10
tangerines (*gut jai*), 17, 20, 25, 37
tang yuan (sweet rice flour dumplings), 26,
    30
Taoism, Taoists, 62, 66, 73, 89, 89, 201
taro, 100
tea (*cha*), 47, 102, 104, 110, 125, 230–34
    brewing of, 233–34
    ceremony, 233–34
    varieties of, 231–33
    in wedding ceremonies, 135, 138
tea ceremony, 233–34
Temple of Heaven, 115
thanksgiving, 95, 104
thank-you notes, 220
Tieguanyin, 232
tiger(s), 165
    year of, 35
time, 178
tin cans, 56
toad, 63
toasting, 140, 177, 220, 229
tofu, 202, 206

tong yuen, 127
toothpicks, 221
Toronto International Dragon Boat Race
   Festival, 75
tortoise, 184
Tray of Togetherness, 19, 129
"treasure blue," 175
Tree of Immortality, 98
Tsingtao beer, 230

umbrellas, 219
United States Dragon Boat Federation,
   75

vegetarianism, 89

wailers, 202
waitstaff, 226
wakes, 201–2
walnuts, 19
Wan Bo, 51–52
wardrobe, 13–14
Warring States Period, 53, 64
water, 56, 124
watermelon seeds, 102, 104
Wealth, God of, 11
Weaving Maiden, 80–82
weddings, 119–50
   attire for, 129–31
   banquet, 138–41
   cuisine for, 140–41
   financial responsibilities in, 139
   five rituals of, 134–38
   jewelry for, 131–32
   night, 142
   planning guide for, 143–50
Wei Chi Jingde, 10
Wen, Duke of, 53–54
Western traditions, 122, 130, 132, 137,
   139, 179, 191, 197, 202–3

white:
   death symbolized by, 12, 13, 31, 130,
      192, 196, 197, 200, 210, 219
   envelopes, 200–201
White Peony, 233
White Rabbit, 97
white tea, 233
widow, 197–98
Wild Ginger (Min), 162
Willow, 45, 54–55
wines, 47, 110, 137, 177
winter solstice, xvi
Woman Warrior, The (Kingston), 86
wood, 124
woodcutter, 98
World Cup Crew Championships, 73
World Dragon Boat Championships, 73
Wu Gang, 98

yams, 29
yang, 11, 48, 62–63, 114, 115, 122, 154,
   162, 206, 218
Yangzi River, 65
yeet-hay (hot breath), 154
yellow, 192, 196, 219
   ribbons, 52
yin, 11, 48, 62–63, 96, 97, 115, 122,
   154, 162
Young, Ruby, 156
Yuan dynasty, 104
Yuan Xiao, 30
Yulan prayers, 90
Yulan Temple Ceremony, 89
yummy peanuts, 103

zang (jewel of pure thought), 91
Zhang, 7
Zhong Kui, 63
Zhucha, 232
zodiac, xvi, 33–40, 123–24, 162, 165,
   171

Photo credit: Joyce Yokomizo

Rosemary Gong is a third-generation Chinese American. Born and raised in California, she and her family were the only Asians in their Central Valley town, school, and church. A graduate of San Jose State University's School of Journalism and Mass Communications, she moved to San Francisco to join Saatchi and Saatchi Corporate Communications Group. Chinatown was her choice of residence—just around the corner from her grandparents. There, she developed an affinity for the places and the customs inherent in her culture. Rosemary shamefully admits her Chinese vocabulary totals around twenty-nine words. Thus, she considers one of her greatest achievements to be ordering a plate of noodles in her native Cantonese.

www.goodlucklife.com